MATH for
merchandising

A Step-by-Step Approach

Evelyn C. Moore

Prentice Hall

Upper Saddle River, New Jersey • Columbus, Ohio

Library of Congress Cataloging-in-Publication Data

Moore, Evelyn C.
    Math for merchandising : a step-by-step approach / Evelyn C. Moore.
        p.   cm.
    Includes index.
    ISBN 0-13-268723-2 (pbk.)
    1. Retail trade—Mathematics.   2. Retail trade—Mathematics—
Problems, exercises, etc.   I. Title.
    HF5695.5.R45M66   1998
    658.8'7'001513—dc21   97-25438
                                                                                     CIP

Editor: Stephen Helba
Production Editor: Patricia S. Kelly
Copyeditor: Linda Thompson
Design Coordinator: Karrie M. Converse
Text Designer: Rebecca M. Bobb
Cover Designer: Russ Maselli
Production Manager: Laura Messerly
Illustrations: Dartmouth Publishing
Marketing Manager: Danny Hoyt

This book was set in Sabon and Formata Condensed by Carlisle Communications, Ltd. and was printed and bound by Banta Company. The cover was printed by Phoenix Color Corp.

© 1998 by Prentice-Hall, Inc.
Upper Saddle River, New Jersey 07458

10 9 8 7 6 5 4 3 2

ISBN: 0-13-268723-2

Prentice-Hall International (UK) Limited, *London*
Prentice-Hall of Australia Pty. Limited, *Sydney*
Prentice-Hall of Canada, Inc., *Toronto*
Prentice-Hall Hispanoamericana, S. A., *Mexico*
Prentice-Hall of India Private Limited, *New Delhi*
Prentice-Hall of Japan, Inc., *Tokyo*
Simon & Schuster Asia Pte. Ltd., *Singapore*
Editora Prentice-Hall do Brasil, Ltda., *Rio de Janeiro*

Dedicated to
Jesse, Peter, and Zachary

preface

The mere thought of taking a math course causes most people to clench their teeth, break out in a cold sweat, and start biting their fingernails. Relax! This course is different.

This course uses practical applications to help you understand the tools of the trade. The approach is geared to help you interpret industry words and thoughts and then use your calculators (or computers) to translate your needs into clear mathematical answers.

You will approach this course in a very logical manner, with a step-by-step approach, one that parallels your career path in the merchandising industry. From the start in Chapter 1, you will discover with the help of the text, which uses a worktext format, that your calculator is a key tool for solving problems effectively.

Chapter 2 teaches you the fundamentals of working with numbers. You look at the relationship of whole numbers to parts so you can calculate sales figures, commission statements, taxes, and discounts. With the numbers serving as the foundation, you can then look at how the numbers reflect the consumer, economic, fashion, and lifestyle trends that businesses address daily.

Once you grasp working with numbers, the work will flow, just as though you were on the job, to more responsible tasks. In Chapter 3 you will look at some of the forms you may be asked to complete in a clerical position or as an assistant buyer. Along with the forms you will learn what you will be filling in, and why. The information on these forms comes from a buyer's purchases at market. You'll take an inside look at the buyer's role in the marketplace, as he or she must negotiate prices with the wholesalers to arrive at the sharpest terms and conditions of sale, including product price, payment arrangements, and shipping charges.

The text then takes you to the retail end of merchandising pricing and repricing products. In Chapters 4 and 5 you apply the basic math skills you learned in Chapter 2 to determine individual, initial, average, cumulative, and maintained markups. Through the exercises in Chapter 5, you continue to develop strong critical thinking skills that reinforce pricing decisions. Markdowns, a very strong component in the competitive retailing world, are covered in Chapter 6.

As you move on in the text, you see how job responsibilities expand and provide further challenges. Part IV of the workbook is designed to help merchandising majors learn the financial planning methods used in the industry. This section covers six-month plans, open to buy, and classification planning. Chapter 7 introduces you to the elements of a six-month plan and explains why they are important to a merchandising operation. From there you move on to Chapter 8, where you learn how to analyze and interpret what the numbers mean and how a merchant can use these figures to judge the overall "healthiness" of an operation. Chapters 9 and 10 carry you to a different level, that of the planner. With a solid foundation in analyzing numbers, adding on markup, and applying markdown pricing, as a merchandiser you now plan stocks, balance the flow of new merchandise and maintain balanced stocks, first by using last year's figures as a guide in Chapter 9 and, then, in Chapter 10, by designing a plan from scratch, just as you would do for a new business. Chapter 11 helps you prepare buying plans for market, which are then reinforced in Chapter 12 as you learn how to build strong merchandise assortments through classification planning.

Part V shows you how numbers serve as tools to use in determining if a company's objectives and goals have been met. Here you take a look at how buying, pricing, and planning decisions are measured and evaluated. Again, using the skills from Chapter 2, you apply basic math skills to profit-and-loss statements and income statements in Chapter 13. Sales per square foot, a key factor in profitability, is introduced in Chapter 14.

Part VI briefly introduces the basics of corporate buying offices. With an increase in national brand products and private labeling growing worldwide, merchandisers faced with increasing competition now have to be able to calculate the cost of goods sold *and* determine if it is feasible to develop a product for a company. In this chapter you learn how to prepare cost sheets and apply the pricing concepts you learned in Part III to determine if a product is competitive. Here you get a glimpse of how merchandising strategies are developing for the 21st century.

The final section provides a check-in point for students. Often students want to make sure they are doing the calculations correctly, but if they are working outside the classroom, they don't have anyone with whom to check. Basic formulas and the solutions to the odd-numbered problems are given.

So, relax! You will take this course step by step, just like your career in the industry. This text will give you the big picture, serving as a "reality check" for what really goes on behind the store windows.

Hands-on experience is always the first step in on-the-job training, and this is a great place for all of you to start. *Math for Merchandising: A Step-by-Step Approach* guides you through the common-sense steps needed as you develop visionary ideas, forecast trends, and end up with financial success in the ever-changing fashion merchandising world.

Acknowledgments

Completion of this project was due in great part to my students, who, for many years, have challenged me to find better and easier ways to teach them the merchandising math skills needed for success in the job market. I am grateful for their insistence and their one constantly repeated question, What do I do first? I thank all of you for reading and improving the materials in this manuscript over the years, but, most importantly, for the confidence you've placed in me.

Many people at Prentice Hall have played significant roles in the completion of this project, and I wish to extend my special thanks to Mark Cohen for his encouragement and support and to Stephen Helba and Patty Kelly for their guidance, judgment, and patience.

And, most importantly, I would like to say to copyeditor Linda Thompson: Your advice, suggestions, and expertise have been invaluable to me, and I honestly cannot begin to thank you enough.

Evelyn Moore

contents

Introduction

one

Using Your Workbook and Calculator

Using Your Workbook

This is a working manual. The best way to learn and apply new math skills is to work with them, so this text has been designed with that format in mind. Each section has

- Key industry terms
- Reasons for doing the work that you are given, and applications of these skills to the job
- Helpful hints
- Reminders
- Examples of forms: how they are filled in and calculated

Grab a pencil, write your own notes in the book, and become involved in learning the math skills you'll need to be successful in the merchandising industry.

Consider this workbook as a new pair of running shoes. At first your shoes feel stiff and uncomfortable, but the more you wear them and the more you exercise, the sooner you will forget that you even have new shoes on, because they are comfortable and feel like they are a part of you.

The same thoughts can be applied to this workbook. Some of the material may feel a little stiff at first. However, if you work through it and then go back and reread the text and problems, the workbook may look worse for wear, but the concepts and critical thinking skills you will come to understand will give you a stronger grasp of the financial operations you need to be successful in the retail merchandising industry.

Using Your Calculator

Knowing how to use tools correctly when working in the business world is one of the biggest keys to success. So now it's time to talk about the different keys on a calculator and how you can use them to solve problems quickly and efficiently.

First of all, it is important to know that all calculators have been programmed differently, technology continues to change, and every calculator you use will be different. When you buy a calculator, it is a good idea to go through the instructions, but most of us don't. In fact, you probably have a calculator at home that you can use for this course, but it is unlikely that the directions are still hanging around. So, let's take a look at the keys and features that are common to most brands.

Common Operating Keys

On/CA   Turns on the calculator; if numbers are showing, touching this key clears everything.

C   Clears the current display.

CE   Clears the last entry without wiping out columns of figures (best time saver in the world).

CA   Clears everything.

C/CE   One touch clears the last entry and two touches clear out all transactions.

Basic Keys

Calculators have four basic keys: addition (+), subtraction (–), multiplication (×), and division (÷). These four operations can be used continuously and can be combined, as long as you don't press the equal (=) key. Pressing the equal (=) key will stop all operations and give a total.

Decimal Point

When you press a digit from 0 to 9, the decimal remains at the right of the entry. If you press the decimal point key, you will key in a decimal fraction, and the decimal point will move to the left. For example, if you press 1 ⊞ 2 ⊞ 5 ⊞ the calculator will show 8. To add 50 cents, you press ⊞, the decimal point, and 50. The display will show 8.50. The decimal point remained at the right of the entry until you told the machine differently.

Percent Key

You should plan to make the percent (%) key your best friend. This key can assist you in many operations, but you have to know how it is going to work on your calculator. The easiest way to see how the % key works is to try it out.

Let's walk through an illustration:

What is $42\frac{1}{2}$% of 200?

Enter 200.

Press the ☒ .

Press 42, the decimal point, and 5 (because $42\frac{1}{2} = 42.5$).

Now press ⊠ .

If 85 appears without touching ⊟ , you have found $42\frac{1}{2}$% of 200.

If your calculator shows 8,500 or if it does not have a % key, your calculator does not compute percents automatically. You may need to touch ⊟ to get 85. If that does not work, then you will have to do the percentage calculations by writing the percent as a decimal and then multiplying. That is, you will have to multiply 200 times 0.425.

If you have a calculator that doesn't do percentages, invest in an inexpensive calculator that has the percentage function programmed in, because such calculators will automatically calculate percentage add-ons or deductions and are the most practical for merchandising math problems.

Let's do another example.

E X A M P L E

A shirt sells for $55. The sales tax in Broward County, Florida, is 6%. Find the total cost of the shirt.

Follow these steps:

Enter 55.

Press ⊞ .

Press 6.

Press ⊠ .

The display should show 58.3 ($58.30), which is $55 plus 6% of $55. This feature is called an *automatic add-on*. You didn't have to find $55 \times 0.06 = 3.30$ and then add 55 to it; that step was done automatically for you.

You can calculate discounts or deductions in the same way. ∎

E X A M P L E

A designer is selling dresses for 30% off the $200 ticket price. Find the selling price.

Follow these steps:

Enter 200
Press ⬚ .
Press 30.
Press ⬚ .

The display should read 140, which is $200 minus 30% of $200.

This feature is called an *automatic deduction*. You didn't have to find $200 \times 0.30 = 60$ and then subtract from 200 to determine the correct answer. ∎

In these examples, the work was done by the tools. You simply had to think through what you wanted the machine to do. Don't forget to use common sense, however, When you look at the display, always ask yourself, Does this answer seem logical?

two

Percentages

Using Percents

When you first enter the fashion-merchandising world, you are swept up in a whirlwind that introduces you to the leading edge of lifestyle and design trends. You discover exciting businesses driven by visionary owners, managers, artists, and designers in search of new horizons. These first introductions generally come as you work in sales or merchandising positions. In these areas you learn about the philosophies of the businesses, which lead designers, manufacturers, and—most importantly—consumers. You learn the importance of following and interpreting economic trends to understand why products fit certain consumer groups so you can make decisions reflecting the wants, needs, and buying patterns of target markets.

Once a business plan is in place, you need to be able to judge how successfully the business is operating. Numbers are the facts in financial analysis. Business owners rely on numbers as the tools needed to analyze and implement changes to expand and develop their businesses. And, to be successful, you quickly discover a solid understanding of basic math is an essential and crucial skill in making the right decisions.

First, you need to start working with numbers as tools. Then, it is important to understand the steps you need to follow to find the answers you are looking for—and to know why you need to follow them. In the merchandising industry you speak in dollars, fractions, and percentages. But, do you know how those numbers were determined or how you have to apply them? When you have a clear understanding of how numbers and terms interrelate, you will create a visual picture that shows if a business is struggling, maintaining itself, or growing. In this chapter, you will look at numbers as wholes and parts and then develop an understanding of how percentages are used.

Percentages are used daily in merchandising for the following purposes.

Comparisons For example, percents are used to compare sales from year to year, compare the amount of stock in different stores, compare how much of the retail price is going to cover expenses or profit, or even to determine how much of the retail selling price goes to the vendor to pay for the merchandise that is sold.

Discounts For example, percents can be used to show the discounts that merchants are given in buying merchandise at a wholesale level for the stores or that consumers are given because of point-of-purchase promotions.

Taxes For example, percents are used in calculating the amount of tax that is due.

Commissions For example, percents are used to determine the amount of payment a salesperson receives if his or her earnings are determined by the amount of sales generated.

You will also find that percentages are used to calculate markup, markdowns, and sale projections and to compare the elements of profit and loss statements.

Finding percentages is simple, because a calculator actually does the work for you. But, it is important to understand the use of percentages and how the numbers are calculated. Even if you are not the person putting the numbers together or the one responsible for the financial decisions, you need to understand how the numbers are determined and to be able to verify such numbers. Developing this understanding is what this chapter is all about.

Basic Math Skills

Let's take a look at some basic general math terms and related skills.

Percentage An amount determined by taking a part, or percent, of another number, the base.

Rate The percent, or the ratio of a number to 100; the symbol representing rate is %.

Base The whole, or total, amount.

Decimals Numbers are written as whole numbers and parts of whole numbers using a decimal point and tenths, hundredths, thousandths, ten thousandths, and so on.

Decimal Point A symbol that separates a whole number from the decimal part.

Fraction A number that names part of a set or part of a whole.

Rounding Replacing a number with one that is easier to use.

Working with Fractions

Fractions consist of two parts. The *denominator* of a fraction tells the number of equal parts, and the *numerator* identifies how many parts are being considered. In merchandising we speak in fractional terms all the time, but when we are working out problems, we have to write the decimal equivalent of the fraction. For example, a buyer might purchase $\frac{1}{2}$ dozen of a product, or take a $\frac{1}{3}$-off markdown. However, most calculators don't have keys for working with fractions, so you must be able to write the decimal equivalents of fractions so you can solve your merchandising problems.

E X A M P L E

The manager of Tom's Sporting Goods wants to divide new inventory among several stores and send $\frac{1}{5}$ of the 620 pairs of new jogging shorts to one store. How many pairs will go to the store?

First, write the decimal equivalent of $\frac{1}{5}$.

$$1 \div 5 = 0.20$$

Now multiply 620×0.20 to determine the answer.

$$620 \boxed{\times} .20 \boxed{=} \quad \boxed{124}$$

The store will get 124 pairs of jogging shorts.

You can also use a calculator to multiply 75 by $20\frac{1}{4}$. First, write $20\frac{1}{4}$ as a decimal. Because $1 \div 4 = 0.25$, you can enter 20.25 instead of $20\frac{1}{4}$. Multiply 75×20.25 to determine the answer, 1,518.75

Rounding

When doing calculations in business, it is often necessary to round the results of the calculations. The following procedure and example illustrate how to round.

First, find the place to which you want to round. In merchandising you will generally round to two places to the right of the decimal point, or the cents place. To round to the nearest cent, you begin by looking at the third decimal place. If that digit is 5 or greater, you add 1 to the second digit following the decimal point. If the third digit is less than 5, leave the second digit unchanged. You drop the remaining digits to the right of the cents place.

E X A M P L E

$$\$36.874 = \$36.87$$
$$\$73.975 = \$73.98$$

10

You follow the same procedure to round to one place after the decimal:

Examples: 87,902.73 = 87,902.7

               2,638.762 = 2,638.8

Dozens

When you are working with basic merchandising problems, you will frequently work with fractions, because many manufacturers package products by the dozens. The following charts show dozens and fractions of dozens.

     1 dozen =   12 pieces
     2 dozen =   24 pieces
     3 dozen =   36 pieces
     4 dozen =   48 pieces
     5 dozen =   60 pieces
     6 dozen =   72 pieces
     7 dozen =   84 pieces
     8 dozen =   96 pieces
     9 dozen = 108 pieces
    10 dozen = 120 pieces
    11 dozen = 132 pieces
    12 dozen = 144 pieces (also known as a *gross*)

$\frac{1}{3}$ is 4 pieces $\frac{1}{6}$ is 2 pieces $\frac{2}{3}$ is 8 pieces

$\frac{1}{4}$ is 3 pieces $\frac{1}{2}$ is 6 pieces $\frac{3}{4}$ is 9 pieces

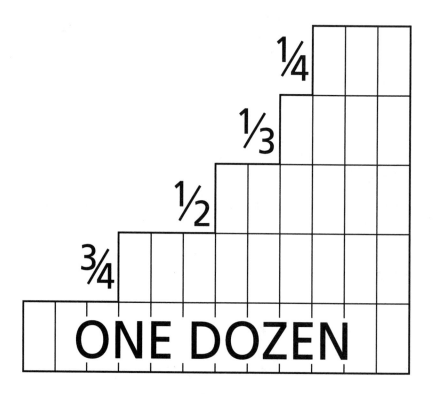

Remember: It is important to know the values of dozens, because we often speak in terms of dozens and purchase in dozens, but we sell in pieces, so we must always make the appropriate conversions.

EXAMPLE

$14\frac{3}{4}$ dozen zippers were shipped back to a supplier because they were the wrong length for the men's trousers. How many zippers were returned?

$14 \times 12 = 168$ zippers.

$\frac{3}{4}$ of a dozen is 9 pieces.

$168 + 9 = 177$ pieces were returned.

Using The Percentage Formula

In this chapter you will apply these basic math terms and skills to the business world. As you learn to analyze whether or not firms are successful, you will look at the **total amounts** (called the **base**) and see how **specific parts** (the **percentages**) of business operations compare by looking at the **percent** (or **rate**).

General Math Terms	Business Industry Terms
Rate	Percent (the % sign means part of 100)
Base	Total amount
Percentage	Specific amount

In general math you learned the following formulas:

Base $\times$ rate = percentage

Rate = percentage $\div$ base

Base = percentage $\div$ rate

What you learned in general math still applies, but you will find more user-friendly terms in the business world. In merchandising you are looking at *whole* numbers: total sales, total pieces. You know that you need to understand how all the *parts* involved fit together. To keep it simple, you will compare and evaluate numbers in the following ways so you can work quickly and efficiently.

Total amount × percent = specific amount

Specific amount ÷ total amount = percent

Specific amount ÷ percent = total amount

There are several important new terms, which we introduce in each chapter.

12

Industry **Terms** and *jargon*

LY Last year.

TY This year.

K Thousands.

Gross Sales All the sales made in a specific period of time.

Customer Returns and Allowances Credits issued for merchandise brought back after a sale or a credit allowed for a defect or price adjustment (allowance).

Volume or Net Sales Total final sales for a period of time.

The following example illustrates how these terms are applied.

Gross sales	$318,000	TY gross sales	
− Customer returns and allowance	− 62,000	CR/A	
= Net sales	= $256,000	Net sales TY (= 100%)	

Helpful Hints

Now it's time for you to start working problems. Look over the next few pages very carefully. In these pages you will learn four basic *Helpful Hints* to make your work easier. That's right, only four . . . and you will use these hints as the foundation throughout this workbook. But, even more importantly, these *Helpful Hints* will be important to you throughout your career. You will use these hints to calculate daily sales projections, as the foundation for calculating markup, to determine the monthly sales performance in a 6 month plan, or even to find the operating expenses in a profit or loss statement.

So, grab a pencil and your calculator, go through the examples, and use the *Helpful Hints* to solve the problems in this chapter.

Helpful hint #1
How to Find Specific Amounts

Use this hint when you are trying to find the specific amount:

- You know the total amount and the percent.
- You want to determine the specific amount.

Example

Taxes reduce your paycheck each week by 22%. You earn $275.00 per week. How much (specific amount) is deducted? What is your net pay?

How to Solve

To find the specific amount when the percent and the total amount
are known, multiply:

$$\text{Total amount} \times \% = \text{specific amount}$$
$$\$275.00 \times 22\% = \$60.50$$

$$
\begin{array}{r}
\$275 \\
\times\ 0.22 \\
\hline
\$60.50
\end{array}
$$

Now simply deduct $60.50 from $275.00 to determine the
net pay.

$$\$275.00 - \$60.50 = \$214.50 \quad \text{net pay}$$

To use the percent key on your calculator,

275 ⊟ 22 %   214.5

Check Your Work

It is important to get into the habit of checking your work.
Remember you are working with wholes and parts to find out if you
solved this problem correctly. Use the business rules on page 11.
For instance, specific amount divided by the total amount gives the
percent.

$$\$60.50 \div \$275.00 = 22\%$$
or
$$\$60.50 \div 22\% \quad = \$275.00$$

or $100\% - 22\% = 78\%$ and $\$275.00 \times 78\% = \$214.50$

Helpful hint # 1:
> **Total amount** $\times$ **percent** = **specific amount**

Helpful hint #2
How to Find the Percent

Use this hint when you are trying to find the percent:

- You know the total amount and the specific amount.
- You want to determine the percent.

Example

Jeans Plus has 40 pairs of pocket jeans in stock. They have 500 pairs
of jeans in the stock overall. What percent of the stock is pocket
jeans?

How to Solve

To find the percent when the specific amount and the total amount are known, divide:

$$0.08 = 8\%$$
$$500 \overline{)40.00}$$

Specific amount ÷ total amount = percent
40 ÷ 500 = 8%

To use your calculator,   40 ⌗ 500 %   8

Check Your Work

Use the formulas on page 11.

$$500 \times 8\% = 40$$
$$40 \div 8\% = 500$$

Remember: Enter the numbers and operations into your calculator exactly as you state them in the problem. The calculator is a great tool, but you must enter the numbers correctly.

Helpful hint #2:
Specific amount divided by total amount = percent

Helpful hint #3
How to Find the Total Amount

Use this hint when you are trying to find the total amount:

- You know the specific amount and the percent.
- You want to determine the total amounts.

Example

To manufacture a blouse, the designer must purchase $6.00 in trims. That $6.00 is equivalent to 8% of the cost of the blouse. What is the cost of the blouse?

How to Solve

To find the total when you know the specific amount and the percent, divide:

$$75 = \$75.00$$
$$0.08 \overline{)6.00}$$

Specific amount ÷ percent = total amount
$6.00 ÷ 8% = $75.00

To use your calculator,

6 ⌗ 8%   75

Check Your Work
Use the formulas on page 11.

$$\$75.00 \times 8\% = \$6.00$$
$$\$6.00 \div \$75.00 = 8\%$$

Note to the Student You can see the solution and the checks all involve the same formulas. Don't be afraid to try these formulas on the job to do some quick calculations involving merchandise or even with your own paychecks or commission statements.

> *Helpful hint #3:*
> **Specific amount divided by percent = total amount**

Helpful hint #4
How to Compare

Use this hint when you are trying to compare, or find the percent of increase or decrease:

- You know an original amount and a new amount.
- You want to find the percent of increase or decrease between the original and new values.

Example
Last year's (LY) sales were $2,000. This year's (TY) sales were $2,500. Find the percent of increase or decrease.

How to Solve
To find the percent of increase or decrease, follow these steps:

1. Subtract the smaller amount from the larger amount.
2. Divide the difference by the original amount

Difference ÷ original amount = percent of increase or decrease

$$\$2,500 - \$2,000 = \$500 \quad \text{difference}$$
$$\$500 \div \$2,000 = 25\% \quad \text{increase}$$

$$\begin{array}{r} 2500 \\ -\ 2000 \\ \hline 500 \end{array}$$

To use your calculator, press

$$2500\ \boxed{-}\ 2000\ \boxed{\div}\ 2000\ \boxed{\%} \qquad \boxed{25}$$

$$\begin{array}{r} .25 = 25\% \\ 2000\overline{)500.00} \end{array}$$

Check Your Work
Use the automatic add-on feature on the calculator:

$$2000\ \boxed{+}\ 25\ \boxed{\%} \qquad \boxed{2500}$$

Or, find the specific amount of increase and add it to the original amount:

$$\$2,000 \times 25\% = 500$$
$$\$2,000 + 500 = \$2,500$$

Note to the Student If this year's sales are greater than last year's, the answer will be a percent of increase. If they are less than last year's, the answer will be a percent of decrease.

Helpful hint #4:
Difference divided by original amount = percent of increase (or decrease)

Summary

An easy way to learn these helpful hints is to always ask the following:

What is the *total amount?*
What is the *specific amount?*
What is the *percent?*

Then, follow these steps:

Fill in what you know.
Identify what is missing.
Review the helpful hints.
Solve the problem.

Remember the basic *Helpful Hints:*

Helpful hint #1:
Total amount × percent = specific amount

Helpful hint #2:
Specific amount ÷ total amount = percent

Helpful hint #3:
Specific amount ÷ percent = total amount

Helpful hint #4:
Difference ÷ original amount = percent of increase (or decrease)

In the problems you will practice how to take advantage of your calculator by using the automatic add-on and subtraction functions. These first problems are designed to help you learn how to

Use the calculator effectively and easily.

Convert dozens and pieces.

Write fractions as percents.

Use *logic* when problem solving.

■ Problems

When solving problems always try to do the following:

- List what you know.
- Identify what you want to determine.
- Determine the helpful hint you need to solve the problem.
- Solve the problem.
- Look at all the numbers. Is the answer logical?
- Check your work.

Percents

1. 26% of $2,573.85 = _____

   1. List what you know: *You know 26% is the percent and $2573.85 is the total amount.*

   2. Identify what you want to determine: *You want to determine the specific amount.*

   3. Use Helpful Hint 1.

   4. Solve the problem. Use your calculator: 2573.85 ⊠ 26%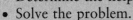
      669.201
      Round to the nearest cent: $669.20.

   5. This answer makes sense: One-fourth of $2,500 is $625.00, so this answer seems reasonable.

   6. Check: $669.20 ÷ $2,573.85 = 25.99%, which rounds to 26%, or $669.20 ÷ 26% = $2,573.85

2. 7% of $800.00 = _____

3. 106% of 872.00 = _____

4. 75 is _____% of 725.50

5. 240 is _____% of 8100

6. 240 is _____% of 810

7. $52.25 is _____% of $1,050.00

Helpful hint #1:
    Total amount × percent = specific amount

Helpful hint #2:
    Specific amount ÷ total amount = percent

Helpful hint #3:
    Specific amount ÷ percent = total amount

Helpful hint #4:
    Difference ÷ original amount = percent of increase (or decrease)

8. $487.00 = 6.5% of _____

9. 55% of 3,650.00 = _____

10. 36.25 = 45.5% of _____

11. $4,275.50 = 39\frac{1}{2}\%$ of _____

(*Tip:* Remember to write $\frac{1}{2}$ as a decimal and then solve for the answer.)

Problems like these are often the most difficult to solve, because you are dealing only with numbers and don't really know how the numbers relate. You will find these situations when you are looking at sales reports by stores, classification summaries, and vendor analysis reports, just to name a few. It will be your job to determine what the relationship is between the whole and the parts and develop a clear understanding of what the numbers represent. By following the *Helpful Hints,* you can *always* find the answer.

12. $4\frac{1}{4}\%$ of 6,000 = _____

13. 400 = 100% of _____

14. 25.75 = 3.5% of _____

15. $62.98 = _____% of $185.95

Helpful hint #1:
      Total amount × percent = specific amount

Helpful hint #2:
      Specific amount ÷ total amount = percent

Helpful hint #3:
      Specific amount ÷ percent = total amount

Helpful hint #4:
  Difference ÷ original amount = percent of increase (or decrease)

16. LY sales = $1,500.00

    TY sales = $1,800.00

    Percent of increase/decrease = _____

17. LY sales = $2,500.00

    TY sales = $2,750.00

    Percent of increase/decrease = _____

18. LY sales = $10,800.00

    TY sales = $9,800.00

    Percent of increase/decrease = _____

19. LY sales = $250K

    TY sales = $315K

    Percent of increase/decrease = _____

    (*Remember:* K means thousands.)

LAST YEAR (LY)							THIS YEAR (TY)						
S	M	T	W	T	F	S	S	M	T	W	T	F	S
						1							
2	3	4	5	6	7	8	1	2	3	4	5	6	7
9	10	11	12	13	14	15	8	9	10	11	12	13	14
16	17	18	19	20	21	22	15	16	17	18	19	20	21
23	24	25	26	27	28	29	22	23	24	25	26	27	28
30							29	30					

20. LY sales = $650,000.00

    TY sales = $585,000.00

    Percent of increase/decrease = _____

21. $2\frac{3}{4}$ dozen lamps were shipped back to the manufacturer because of a defect in the plugs. How many lamps were returned?

22. $24\frac{2}{3}$ dozen weight sets were placed on the floor for Father's Day. How many weight sets were for sale?

23. A buyer purchased the following items. Convert the quantities from dozens to pieces:

$\frac{2}{3}$ dozen sport watches: _____ pieces

$2\frac{1}{2}$ dozen surfer bands: _____ pieces

$3\frac{1}{4}$ dozen water socks: _____ pairs

$5\frac{1}{3}$ dozen sunglasses: _____ pairs

$1\frac{1}{2}$ gross decals: _____ pieces

24. A buyer for a men's sweater department purchased 16 dozen cardigan sweaters for Christmas. The buyer purchased the following:

Small: 2 dozen = _____ number of pieces

Medium: $4\frac{1}{3}$ dozen = _____ number of pieces

Large: $6\frac{2}{3}$ dozen = _____ number of pieces

X-large: 3 dozen = _____ number of pieces

25. Convert the following from dozens to pieces:

$18\frac{2}{3}$ dozen = _____ pieces

$1\frac{1}{2}$ dozen = _____ pieces

$4\frac{3}{4}$ dozen = _____ pieces

$11\frac{1}{3}$ dozen = _____ pieces

$9\frac{2}{12}$ dozen = _____ pieces

$7\frac{1}{2}$ dozen = _____ pieces

$1\frac{1}{4}$ gross = _____ pieces

$20\frac{1}{6}$ dozen = _____ pieces

26. A buyer purchased $3\frac{1}{2}$ dozen T-shirts at $48.00 per dozen.

    How many shirts were purchased? _____

    What was the total cost of the shirts? _____

27. Six dozen pens were placed on sale for $8.99 each. After the sale, $2\frac{1}{2}$ dozen were left.

    How many pens were on sale originally? _____

    How many dozens of pens were sold? _____

    How many pens were left at the end of the sale? _____

    What was the overall sales volume generated on the pens sold? _____

Applications of Percent

The following problems involve calculations merchandisers make on a daily basis. *Remember:* Use your calculator and a lot of common sense.

28. During the fall back-to-school sale last year, 680 leather-bottom backpacks were sold. This year sales declined by almost 26%. How many backpacks were sold this year?

29. At Thor's Jewelers, 60 watches were placed on sale for $19.99. During the sale $\frac{3}{4}$ of the watches were sold. How many watches were left? (*Tip:* Calculate the decimal equivalent of $\frac{3}{4}$ first and then determine the pieces.)

30. Children's undershirts were selling for $1.69 each. If the merchant purchased 1 gross and 75% of the gross was sold, what were the total sales in dollars? (*Tip:* Calculate pieces sold first; then determine the dollar value.)

31. Golf tees were selling at a special price of 2 packages for $3.50. How many packages could you purchase for $15.75?

32. A buyer discovered that about 15% of the sweaters bought in Taiwan were damaged during shipping. The manufacturer had insurance to cover any damaged goods. 38 dozen sweaters were shipped. Approximately how many were damaged?

33. A young men's department received the following shipment:

    $8\frac{1}{3}$ dozen knit polo shirts

    $6\frac{1}{4}$ dozen oversized shirts

    $11\frac{2}{3}$ dozen novelty shirts

    How many shirts did they receive?

34. Last week Rob's commission was $104.16. If he earns 12% commission, what were his total sales?

35. Cosmetic sales in May were $426,000.00. Maria sold 12% of the total sales. The sales commission for cosmetic sales is 6% of sales. What is Maria's commission check?

36. The sales for the swimwear department last year were $96,000.00. The manager for the department plans an increase of 8.5% for this year. What is this year's projected sales volume? (*Reminder:* Don't forget to use your calculator effectively.)

37. A buyer purchased $20\frac{2}{3}$ dozen bath sheets for a special 3-day sale. 25% of them were monogrammed with a single initial. How many monogrammed bath sheets did the buyer purchase?

38. A sales associate for a specialty CD store earns 18% of the selling price of CDs. He earns $2.52 on each CD. What is the selling price of the CDs?

39. A sales representative for a men's tie company earns a salary of $1,200.00 per month plus a commission based on her total sales. Last month she earned $1,560.00 for the month. Her sales were $7,200.00 What was the rate of her commission?

40. A local boutique negotiates with a designer to sell her one-of-a-kind beach shirts. The owner receives $13.44 from each shirt and the designer receives $14.56. What percent does the store owner receive?

41. The sweater department ran a sale last week and sold 95% of the sweaters that were on sale. 38 sweaters were sold. How many sweaters were on sale?

Helpful hint #1:
Total amount × percent = specific amount

Helpful hint #2:
Specific amount ÷ total amount = percent

Helpful hint #3:
Specific amount ÷ percent = total amount

Helpful hint #4:
Difference ÷ original amount = percent of increase (or decrease)

42. The sales supervisor reviewing the monthly operations expenses determined the following:

70% of the budget was directed for salaries.
10% was used for maintenance and repairs.
5% was used for department supplies.

The remaining monies were earmarked for miscellaneous expenses.

The budget is $6,800.00 for the month. What are the dollar values of these categories?

Salaries _____ Maintenance and repairs _____

Supplies _____ Miscellaneous _____

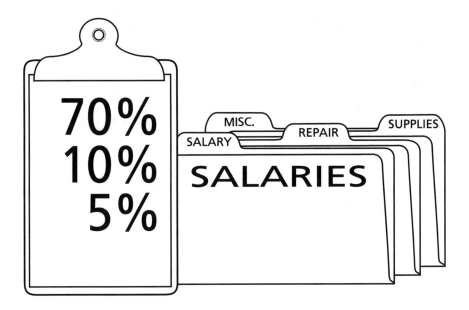

43. Last year the housewares department sold $260,000.00 of pottery cookware. This year the manager plans to sell $364,000.00. What is the percent of increase for this year's sales?

44. A leather goods manufacturer decided to reduce the production of saddlebag-style purses from 300 per day to 225 per day. What was the percent of reduction?

45. Sales this year are off by 20% in the coat department. Last year in October they sold 415 coats. How many coats did they sell this year in October?

46. The volume in the men's tie department last year was $62,000.00. An increase of 15% is projected this year. What is the projected volume this year?

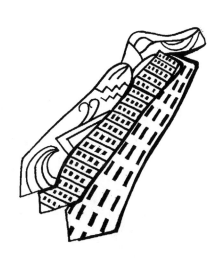

47. Sales of decorative throw pillows are $16,500.00 at the end of the year in the linen department. This represents 4% of the total department sales. What are the total sales for the linen department?

48. A buyer totaled the sales last week and determined that Ellen sold $650.00, Aaron sold $1,100.00, and Sun-li sold $925.00. What was the percent of sales for each sales associate?

Ellen _____ Aaron _____ Sun-li _____

49. A linen department had an inventory of $152,000.00 on July 1. On July 15 the buyer wanted to move out old merchandise to prepare for the August white sale. The following were transferred to an outlet branch:

24 twin sheet sets at $28.00 each

19 blankets at $32.00 each

26 shower curtains $15.00 each

$4\frac{1}{2}$ dozen boxes of shower curtain hooks at $8.00 each

What percent of the inventory was transferred?

50. In the Holiday Shop the manager wants 20% of the total inventory in the stockroom and the rest displayed on the floor. After meeting these instructions, you placed $35,000.00 of the inventory in the stockroom. What is the dollar amount of the inventory on the selling floor?

Total inventory _____

Inventory in the stockroom _____

Inventory on the selling floor _____

51. The merchandise manager projected a 28% increase in business for the young men's 8–20 area this year. The planned sales are $615,000.00 for the fall season. What were last year's sales? (*Hint:* Remember, the amount for last year is 100%, so last year = 100% and this year = 100 + 28%, or $615K (K means thousands). Don't forget to check your work.)

52. A manufacturer projects a 26% increase this year over last year's sales. Last year's sales volume was $104,000.00. Last year's sales were 10% greater than the year before.

a. What are the sales for: 2 years ago _____

Last year _____

This year _____

b. What is the percentage difference between TY and 2 years ago?

Use your *Helpful Hints.*

Helpful hint #4:
Difference ÷ original amount = percent of increase (or decrease)

53. The total sales for the shoe department were $21,000.00 for the first week of May. During the week Janelle worked a total of 16 hours and had sales of $2,100.00. Reuben worked on Saturday for 8 hours and had sales of $1,500.00. Omar was the primary sales associate, putting in 32 hours with sales of $2,380.00.

a. What was each salesperson's percent of sales?

Percent of Sales

Janelle _____

Reuben_____

Omar _____

b. How many dollars per hour did each associate earn? Simply divide sales earned by hours worked.

Dollars Per Hour

Janelle _____

Reuben_____

Omar _____

Give your opinion about the productivity of each of these employees. This example shows how numbers can be used as one tool in merchandising and management decisions.

54. The manager in a sporting goods store reduced a workout bench from $650.00 to $390.00. Determine the percent of reduction.

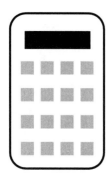

55. A toy department sold 550 action figures last year. This year sales declined by 60%. How many action figures were sold this year?

56. Home furnishing sales are currently on the rise, and the linen department of a leading department store projected a 32% increase in business for this year. This year's projected sales are $325,000.00. What were last year's sales.

57. The volume in the men's hat department last year was $30,000.00. An increase of 28% is projected for this year. What is the projected volume for this year?

58. Last year the small-electronics department sold $175,000.00 in hand-held electronic games. This year the manager plans sales of $105,000.00. What is the percent of increase or decrease for this year's sales? (Use the correct vocabulary in your answer.)

59. The volume in the furniture department last year was $750,000. The manager for that department is planning a 14% decrease for this year. What is this year's projected volume?

60. McKee Company purchased 468 shopping bags in July for their retail store, which was 30% more than last year.

    a. How many dozen bags did they purchase this year? _____

    b. How many bags did they purchase last year? Show your answer in pieces and dozens.

61. The merchandising designers found fabric on an overseas trip to produce new T-shirts for spring. After negotiations the designers found the cost of the fabric to be $1.88 per yard. It takes $\frac{3}{4}$ yard to make one T-shirt. After the buttons and trim are added, the designers determine that the materials for one shirt cost $2.50. What percent of the cost is fabric?

62. Last year the Natural Fabric Manufacturer had gross sales of $102,000.00. Due to a weaving problem with the looms and poor quality control, the company had returns of $34,000.00. This year the firm corrected the problems and is projecting a 32% increase over last year's sales.

    a. What was last year's sales volume? _____

    b. What are this year's projected sales? _____

Percentages and Calculators

As you continue to work with percentages, you will recognize that they are a means to do the following:

1. **Analyze,** thereby providing a statement for a business to use in making an intelligent change in strategy or policy.

2. **Exchange** data and **compare** to determine weakness and strengths.

3. **Provide direction** for a business.

By completing the following problems you will demonstrate that you are comfortable working with your calculator, with totals, and with parts.

63. The sales volume for a local store last year was $30,000,000.00. The hosiery department sales were 1.9% of the total sales. This was $5,700.00 better than the previous year. The handbag department accounted for 2.1% of the total sales last year. Handbags, however, were down $14,000 overall from the previous year. Identify the following:

LY net sales for hosiery _____   Sales 2 years ago _____

LY net sales for handbags _____   Sales 2 years ago _____

64. The Pottery Shop opened a new branch this year. The following figures show the sales volume and advertising expenses.

Sales LY	$ 900,000.00
Sales TY	$1,050,000.00
Sales plan for next year	$ 975,000.00
Advertising costs LY	$ 28,500.00
Advertising costs TY	$ 32,000.00
Advertising plan	$ 27,500.00

a. What do the sales trends reflect?

b. What do the advertising costs reflect?

c. What is the percent of increase in sales from last year to next year's plan?

d. What is the percent of increase or decrease in advertising between last year and next year?

e. What is the percent of decrease in sales from this year to next year?

f. Why do you think these fluctuations of volume and expenses might be planned?

65. Four years ago the Breaker's Kite Company opened and generated a first year's sales volume of $12,000.00. Sales continued to improve, and this year the company projects an 18% increase over last year, bringing the overall sales volume to $62,000.00.

a. What was last year's sales volume?

b. What is the overall percent of increase between this year's sales projections and the original $12,000.00 volume?

PART II

Purchasing Concepts

CHAPTER THREE
Purchasing Products: The Terms,
Conditions, and Forms Used for Buying
Merchandise

three

Purchasing Products: The Terms, Conditions, and Forms Used for Buying Merchandise

As a merchandiser you learn right away that getting the *right merchandise* for the *right price* at the *right time* is truly an art, and anything you can negotiate to reduce costs will help you to build profits. Once you negotiate the best terms for buying your products, you will complete a purchase order. This purchase order is a binding, legal, contractual agreement with explicit conditions to be met by both the buyer and the seller.

As you take on more responsibility in your career, you may find yourself working as a clerical, assistant, or associate buyer for a retailing firm or in the production or purchasing department of a manufacturing company. In these positions, one of the key job responsibilities is preparing orders for the buyers. You not only need to know how to calculate the totals correctly, but also to understand *why* the orders are completed in the way they are.

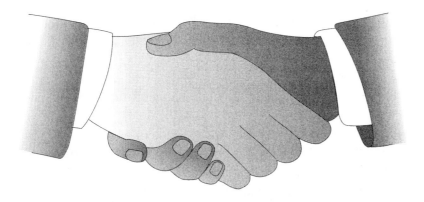

Terms and Conditions

Let's take a look at the key points in an order. Buyers and vendors must agree on a variety of things before an order can be written.

Price of the Product

The vendor wants to sell merchandise at the highest price, but the buyer wants the lowest price. The quoted cost price can often be reduced by applying any one or some combinations of the following purchasing discounts:

trade, quantity, seasonal, and promotional (advertising)

Payment Terms

Payment terms can also be called **credit terms.** The buyer negotiates a cash discount for early payment. Negotiations are twofold:

First, the buyer and seller have to agree on what percentage of cash discount will be offered if the bill is paid promptly.
Second, they must agree on a specific payment date.

Shipping Terms

Simply put, the buyer and seller have to agree on how the merchandise will be shipped and who will pay the shipping charges.

The key is to blend the best of all three conditions and provide the buyer with the best product at the best price at the right time for the greatest profits.

Industry **Terms** and *jargon*

Purchase Order   A contractual agreement outlining price and quantity of product and shipping and payment terms.

Invoice   A bill that outlines specifically the quantity of items that were shipped, the date, and how the merchandise was shipped. Invoices are prepared at the time the merchandise is packed and shipped.

Cost Price   The price at which the manufacturer sells the product to the retailer. Sometimes this is called the **wholesale** price or, simply, the *cost*.

Retail Price   The price the consumer pays.

Wholesale   Often known as the *cost price,* it is the price that is quoted to the retailer.

List Price (MSRP)   Also known as the manufacturer's suggested retail price, a price established by the manufacturer for the consumer so there will be consistency and credibility in pricing. When the manufacturer quotes a list price to the retailer, the cost is determined by deducting trade discounts.

Discount A deduction taken from the original price, usually offered as a percentage. A few of the most common types of discounts offered in the merchandising industry are defined here.

Trade Discount A discount deducted from the list price or cost for both wholesalers and retailers. It is a discount for being in the same business, or trade.

Quantity Discount A discount given for purchasing large amounts in bulk.

Seasonal Discount A discount given for merchandise purchased prior to a normal buying season or after the traditional selling season.

Advertising and Promotional Discount A discount offered to help a merchant advertise or promote merchandise. Often these discounts are based on annual sales volume, and the dollars earned are directly applied to an ad, catalog, or promotional campaign agreed upon by the vendor and the retailer.

Series Discount One or more discounts offered together. Each discount is deducted, *one at a time*, from the preceding amount.

Billed Cost Price determined after all trade, quantity, seasonal, or advertising and promotional discounts are deducted, but before cash discounts are deducted.

Delivered Cost Price determined after all trade, quantity, seasonal or advertising and promotional discounts are deducted, including freight charges, but before cash discounts are deducted.

Cash Discounts A discount given for paying the invoices prior to the negotiated due date.

Net No discount.

Dating The practice of setting a specific time for paying the bill. There are several types:

Ordinary/regular/D.O.I.

R.O.G.

Advanced (as of)

Extra

EOM

COD

These terms are each fully defined, with examples, in the workbook.

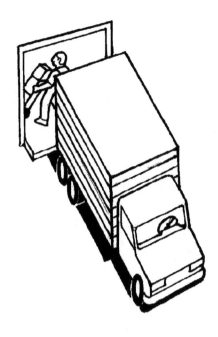

FOB *Free on board* or *freight on board,* a designation that determines when the ownership of goods takes place and, once that exchange occurs, which party is responsible for shipping charges.

General shipping terms are as follows:

FOB store or FOB destination
FOB factory or FOB shipping point
FOB city
FOB 50/50
FOB store prepaid
FOB factory prepaid

These terms are defined, with examples, in the workbook.

Remitted Sent in, or paid.

Unit Price Price per item or unit.

Extended Amount Total value of units or quantity ordered multiplied by the individual price at both cost and retail.

Manufacturer Also called vendor, supplier, or wholesaler, the person or company from whom the store is buying the merchandise or products.

Buyer The store or company buying from the manufacturer or vendor.

Consumer A person purchasing the merchandise in a retail store.

Forms

Now that you've reviewed some of the terms that come up in purchasing, let's take a look at some of the forms you will be using. It is very important to understand the forms to complete the paperwork correctly.

Purchase Orders

On the opposite page is the purchase order, which is a contractual agreement outlining price, quantity, and shipping and payment terms. There are dozens of styles of purchase-order forms in the business world. The layout and design are different on each. The purchase order shown is a typical form, which has the following key components important in negotiating a business agreement.

A. Name of the purchasing company.
B. Company billing address and phone and fax numbers.
C. Company purchase-order number.
D. Vendor name, address, and identification number.
E. Address to which the products are to be shipped.
F. Date of the purchase order.   G. Department name.
H. Department number.   I . Requested shipping date.
J. Cancellation date.
K. Identification of ownership exchange and who pays shipping.
L. Identification of transportation to be used.*
M. Payment terms negotiated.   N. Quantity purchased.
O. Style number.   P. Description.
Q. Size.   R. Color.
S. Cost of one item.   T. Retail price of one item.
U. Total (extended) cost (quantity × unit cost).
V. Total (extended) retail (quantity × unit retail).
W. Total markup dollars (total R$ – total C$ = total MU$). (See Chapter 4.)
X. Retail markup percentage on each item (MU$ ÷ R$). (See Chapter 4.)
Y. Total rows: total number of pieces on the purchase order.
Z. Total cost dollars calculated on the order.
AA. Total retail dollars calculated on the order.
BB. Total markup dollars: total order retail – total order cost. (See Chapter 4.)
CC. Overall retail MU percentage on purchase order: total MU$ ÷ total R$ = total MU%.
DD. The planned departmental markup percentage. (See Chapter 4.)
EE. Where copies are distributed.

---

*It is very important to specify the type of transportation in order to avoid unnecessary freight charges. Don't assume the vendor will ship by truck. If the vendor is late in shipping merchandise, he or she may use overnight shipping services, which are more costly, to meet deadlines if transportation type is not specified.

PURCHASE ORDER

Store Name A

Address

Phone **Fax** B **P.O. no** C

Vendor name D

Vendor address E

Vendor city, state, ZIP

Vendor phone, fax numbers

Vendor no.

Main store and branch store addresses

Order date	F
Department name	G
Department no.	H
Shipping date	I
Cancel if not received by	J
FOB	K
Ship via	L
Terms	M

Ship to

Ship to

Ship to

Order MU % CC

Planned Departmental MU % DD

Qty	Style	Description	Size	Color	Unit Cost	Unit Retail	Total Cost	Total Retail	Total MU $	MU % Each
N	O	P	Q	R	S	T	U	V	W	X
Y						**Total Cost**		**Total Retail**	**Total MU $**	**Overall MU %**
							Z	AA	BB	CC

EE

Totals

Original and 3 copies
1. Original to vendor/manufacturer
2. Copy to store accounting department
3. Copy to store receiving department
4. Copy to buyer's records

Note: The unit retail, total retail, MU $, and MU % are hidden from the manufacturer.

The original shows only the cost values, because that is what the store is paying.

Look at the completed purchase order on page 41.

1. The Tak Room is located at 807 Wembly Road, Los Andreas, CA, where the bills will be sent.
2. The store has three locations to which it wants this merchandise sent. In this case all three stores will receive this same merchandise assortment in the ladies' accessories department, number 650.
3. The date is January 24, and the purchase order number is 4950, which will be referenced on the invoice.
4. The vendor being used is PZ Moderno, which the store will identify in their company records with the vendor identification number 86954.
5. The merchandise is to be shipped on March 20, not before.
6. The merchandise must be received by April 10, or the order will be canceled. If the merchandise is received after April 10, the buyer may return it.
7. The vendor is not paying shipping charges. The exchange of ownership of merchandise occurs when the merchandise leaves the factory. The buyer pays the shipping charges.
8. The merchandise is to be shipped by truck.
9. The payment terms are 8/10 eom, which means an additional 8% discount will be given if the invoice is paid within 10 days from the end of the month.
10. There were 6 pieces of style no. 101 purchased. These scarves came in multicolored patterns and cost $4.50 each. The scarves will be retailed at $10.00 each. The total cost is $27.00 (6 × 4.50) and the total retail value is $60 (6 × $10.00). Each item on the purchase order is extended the same way.

Note that markup is discussed in Chapter 4. The buyer's goal for this department is to achieve a 52% markup for the merchandise purchased. This has been accomplished on this purchase, because the initial markup is 53.96% for the merchandise.

All purchase-order forms are different. Some purchase orders have shaded areas, which usually cover up the retail and markup values the buyer is placing on the merchandise. Or, a purchase order may have a perforation, allowing the retail and markup values to be torn off.

PURCHASE ORDER

Store Name Oak Room

Address 807 Wembly Road Los Andreas, CA 92410
Phone (231) 624-0000   **Fax** (231) 624-5000

P.O. no 4950

Vendor name PZ Moderno

Vendor address 31212 Common Drive

Vendor city, state, ZIP Boston, Ma 02130

Vendor phone, fax numbers (615) 777-2222 (615) 777-2223

Vendor no. 86954

Order date January 24, 19—
Department name Ladies' Accessories
Department no. 650
Shipping date March 20
Cancel if not received by April 10
FOB Factory
Ship via Truck
Terms 8/10 eom

Main store and branch store addresses

Ship to 16 Hillside Road Huntington Beach, CA 92685

Ship to 1716 E Ward Blvd. Western Woods, CA 92750

Ship to 2821 Five Red Road Saylorsburg, CA 92587

Order MU % 53.96%

Planned Departmental MU % 52%

Qty	Style	Description	Size	Color	Unit Cost	Unit Retail	Total Cost	Total Retail	Total MU $	MU % Each
6	101	Print scarves	—	Multi	$4.50 ea	$10.00	$27.00	$60.00	$33.00	55%
6	107	Print scarves	—	Multi	$4.50 ea	$10.00	$27.00	$60.00	$33.00	55%
12	115	Sash belts	—	Multi	$6.00 ea	$12.00	$72.00	$144.00	$72.00	50%
18	136	Chain belts	—	Brass	$6.00 ea	$15.00	$117.00	$270.00	$153.00	56.67%
1/2 doz	142	Hosiery	P/M	Beige	$24.00 doz	$4.00	$12.00	$24.00	$12.00	50%
1/2 doz	142		M	Beige	$24.00 doz	$4.00	$12.00	$24.00	$12.00	50%
1/2 doz	142		L/T	Beige	$24.00 doz	$4.00	$12.00	$24.00	$12.00	50%
Totals							**Total Cost** $279.00	**Total Retail** $606.00	**Total MU $** $327.00	**Overall MU %** 53.96%

Original and 3 copies
1. Original to vendor/manufacturer
2. Copy to store accounting department
3. Copy to store receiving department
4. Copy to buyer's records

Note: The unit retail, total retail, MU $, and MU % are hidden from the manufacturer.

The original shows only the cost values, because that is what the store is paying.

Invoices

An **invoice** is a bill that states the quantity of items shipped, the date, and how the merchandise is shipped. Invoices are prepared at the time the merchandise is packed and shipped. An invoice is prepared for each shipment to each individual store. Generally multiple copies are prepared. The date is very important on the invoice, because it is from that date that accounting departments determine how many days they have to pay the bill.

1. One copy with all the prices goes to the vendor's accounting department.
2. The original copy with all the prices is mailed (or sometimes is attached on the outside of the box) to the purchasing company.
3. A copy identifying styles and quantities accompanies the merchandise for inventory checking when the merchandise is received.
4. A copy goes to the shipping company for value and insurance.

The following information is on an invoice, as illustrated on page 43. This information should agree with what is found on the corresponding purchase order that is noted on the invoice.

A. Name of the vendor.
B. Address and phone and fax numbers of the vendor.
C. Name and address of the store receiving the merchandise.
D. Department name and number.
E. Invoice number.
F. Date the invoice is prepared.
G. Purchase order number from the buyer.
H. Time at which the merchandise changes ownership and who is paying the shipping charges.
I. Method of shipment.
J. Payment terms.
K. Quantity of merchandise shipped.
L. Style number.
M. Description.
N. Size.
O. Color.
P. Cost of each item.
Q. Total cost (quantity × unit cost = total cost).
R. Totals: pieces and dollars.
S. Where the copies are distributed.

INVOICE

Vendor or manufacturer's name

Address   A

City, state, ZIP   B

Phone, fax

Ship to (purchasing store name and address)   C

Department name   D

Invoice no.   E

Date   F

PO no.   G

FOB   H

Ship via   I

Terms   J

No. _____

Qty	Style	Description	Size	Color	Unit Cost	Total Cost
K	L	M	N	O	P	Q
Totals R						

Original and 3 copies
1. Original to manufacturer's accounting department
2. Copy to store with merchandise
3. Copy mailed to store accounting
4. Copy to shipping company   S

The copy that goes to the store accounting office will be mailed to the main store identified on the purchase order.

Note: The invoices will not show a retail value. The retail value is determined by the buyer.

Look at the completed invoice on page 45.

1. PZ Moderno is sending merchandise to the store located at 16 Hillside Road in Huntington Beach, CA. The information on this invoice is only for this store. A copy of this invoice will accompany the merchandise so it can be correctly inventoried, and another copy of the invoice will be mailed to the main store, which is located at 807 Wembly Road, Las Andreas, and identified on the purchase order.
2. The other two stores identified on the purchase order will also receive this merchandise assortment and will be billed accordingly.
3. The invoice number is 514, and it is being prepared on April 6.
4. The purchase order number is 4950, which is also found on the purchase order.
5. The merchandise is shipped on April 6, to the ladies' accessories department, no. 650, in agreement with the purchase order.
6. The vendor is not paying shipping charges, and the exchange of ownership of merchandise occurs when the merchandise leaves the factory.
7. The merchandise is being shipped by truck.
8. The payment terms are 8/10 eom, which means an additional 8% discount will be earned if the invoice is paid within 10 days from the end of the month.
9. Six pieces of style no. 101 were purchased. These scarves came in multicolored patterns and cost $4.50 each. The scarves will be retailed at $10.00 each. The total cost is $27.00 (6 × 4.50) and the total retail value will be $60.00 (6 × $10.00). Each item is calculated in the same manner and reflects the information found on the purchase order.

Return-to-Vendor Forms

Also known as an RTV, charge-back, claim or debit memo, the return-to-vendor form is used when returning merchandise. Reasons for this type of transaction include

Canceled order	Wrong styles	Wrong sizes
Wrong colors	Not ordered	Overshipment
Customer repair	Sample for ad layout	

Sometimes an RTV form is completed if the buyer and vendor have negotiated an advertising agreement for the vendor to contribute money for an advertisement. Probably one of the most important points to know is that you cannot simply send back merchandise or charge back an advertising cost unless you have authorized agreement between the buyer and the seller. It is also important to note on the

Invoice

Vendor or manufacturer's name	PZ Moderno
Address	3121 Common Drive
City, state, ZIP	Boston, MA 02130
Phone, fax	(617) 777-2222 (617) 777-2223

Invoice no.	514
Date	April 6
PO no.	4950
FOB	Factory
Ship via	Truck
Terms	8/10 eom

Ship to (purchasing store name and address):
Tak Room
16 Hillside Road
Huntington Beach, CA 92685

Department name   Ladies' Accessories   No. 650

Qty	Style	Description	Size	Color	Unit Cost	Total Cost
6	101	Printed scarves	–	Multi	$4.50 ea	$27.00
6	107	Printed scarves	–	Multi	$4.50 ea	$27.00
12	115	Sash belts	–	Multi	$6.00 ea	$72.00
18	136	Chain belts	–	Brass	$16.50 ea	$117.00
1/2 doz	142	Hosiery	P/M	Beige	$24.00 doz	$12.00
1/2 doz	142	Hosiery	M	Beige	$24.00 doz	$12.00
1/2 doz	142	Hosiery	L	Beige	$24.00 doz	$12.00
Totals 60						$279.00

Original and 3 copies
1. Original to manufacturer's accounting department
2. Copy to store with merchandise
3. Copy mailed to store accounting
4. Copy to shipping company

The copy that goes to the store accounting office will be mailed to the main store identified on the purchase order.

Note: The invoices will not show a retail value. The retail value is determined by the buyer.

RTV form that retail is never identified to the vendor (seller). An RTV form is shown on page 47. It contains the following information:

A. Store name that bought the merchandise.
B. Address and phone and fax numbers.
C. RTV number.
D. Name and address of the vendor.
E. Vendor identification number.
F. Date merchandise is being returned.
G. Department number.
H. How the merchandise is being shipped.
I. Authorization: person and authorization number.
J. Quantity.   **K.** Style number.
L. Description.   **M.** Size.
N. Color.   **O.** Unit cost.
P. Total cost (quantity × unit cost).
Q. Unit retail.   **R.** Total retail (quantity × unit retail).
S. Reason for return.   **T.** Total pieces returned.
U. Total RTV cost value.   **V.** Total RTV retail value.
W. Where copies are distributed.

The retail price is seen only on the copies that go to the company files and accounting department. The copies received by the shipping company and the vendor (seller) do not show the retail price.

Note: These returns almost always require authorization, which means that PZ Moderno has to approve the return. Sometimes companies will issue an authorization number, and sometimes they identify the person who approved the return.

Look at the completed RTV form for merchandise being returned because it is defective (page 48).

1. The Tak Room is sending merchandise back to PZ Moderno, which is identified by vendor no. 86954, on May 22.
2. Four belts, style no. 136, are being returned because the belts are defective.
3. The RTV form number is 1411.
4. The merchandise is being shipped by truck.
5. The return was authorized by Sarah Shea, with authorization number 22548.
6. The value is $6.50 each, for a total of $26.00 (4 × $6.50 = $26.00).
   The store will be credited for $26.00.

Return-To-Vendor Form

Store name *A* **Fax** *B* **RTV no** *C*

Address
Phone

Return to:

Vendor name   *D*

Address

City, state, ZIP

Vendor no.   *E*

Date   *F*

Department no.   *G*

Ship via   *H*

Authorized by   *I*

Authorization no.

Qty	Style	Description	Size	Color	Unit Cost	Total Cost	Unit Retail	Total Retail	Reason for Return
J	*K*	*L*	*M*	*N*	*O*	*P*	*Q*	*R*	*S*
Totals *T*						*U*		*V*	

Original and 3 copies
1. Original to vendor
2. Copy to store control office
3. Copy with buyer's records
4. Copy attached to merchandise

Note: The unit retail and total retail are hidden from the manufacturer. The original copy shows only the cost values, because it is the cost being credited to the store. The retail is determined by the buyer.

M

RETURN-TO-VENDOR-FORM

Store name _Tak Room_

Address _807 Wembly Road, Los Andreas, CA 92410_

Phone _(213) 624-0000_ **Fax** _(213) 624-5000_

RTV no _1411_

Return to:

Vendor name _PZ Moderno_

Address _31212 Common Drive_

City, state, ZIP _Boston, MA 02130_

Vendor no. _86954_

Date _May 22, 19—_

Department no. _650_

Ship via _Truck_

Authorized by _Sarah Shea_

Authorization no. _22548_

Qty	Style	Description	Size	Color	Unit Cost	Total Cost	Unit Retail	Total Retail	Reason for Return
4	136	Belts	—	—	$6.50	$26.00	$15.00	$60.00	Defective
									buckles

Original and 3 copies
1. Original to vendor
2. Copy to store control office
3. Copy with buyer's records
4. Copy attached to merchandise

Note: The unit retail and total retail are hidden from the manufacturer. The original copy shows only the cost values, because it is the cost being credited to the store. The retail is determined by the buyer.

| **Totals** | 4 | | | | | $26.00 | | | |

Negotiating

Let's get started negotiating.

Discounts

You need to start by negotiating the best price on the raw materials or the finished products. This is done by getting the best discounts. A discount is a deduction from the original price. Let's take a look at the most common discounts and learn how they are calculated.

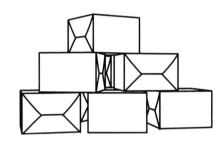

Quantity Discounts **Quantity discounts** are given for purchasing large quantities. The discount may be given on a single order or over a period of time. Fabrics and findings in the manufacturing industry and items such as underwear, hosiery, or consumables are examples of merchandise bought in quantity on the retail end.

E X A M P L E

A manufacturer offers a 6% discount on the entire order when a buyer places an order of 600 pieces or more. An order for 724 pieces was placed at a cost of $6.25 each. How much will the merchandise cost?

   To find the total cost, multiply the number of pieces by the cost per piece and subtract the discount.

$$724 \times \$6.25 = \$4,525.00 \quad \$4,525 \times 6\% = \$271.50$$
$$\$4,525 - \$271.50 = \$4,253.50 \text{ Total cost}$$

To use your calculator, press

$$724 \; \boxed{\times} \; 6.25 \; \boxed{-} \; 6 \; \boxed{\%} \qquad \boxed{4253.5}$$

You can also find the price for each piece with the discount and then multiply by the number of pieces.

$$6.25 \; \boxed{-} \; 6 \; \boxed{\%} \; \boxed{\times} \; 724 = \qquad \boxed{4253.5}$$

The total cost is $4,253.50.

Determine the better buy for the following.

A 15% quantity discount is offered on fabric costing $4.00 per yard if 300 yards are purchased.

If 300 yards are purchased, what is the total cost? _____

What is the cost per yard? _____

A second mill offers similar fabric with no discount at $3.50 per yard.

If 300 yards are purchased, what is the total cost? _____

What is the better buy? _____

Seasonal Discounts **Seasonal discounts** are discounts given for merchandise purchased prior to a normal buying season or after the traditional selling season.

EXAMPLE

A vendor offers bathing suits at the end of a season to a leading retail store. Final negotiations determine that the suits, which originally cost $21.75 each, will be offered with a seasonal discount of 15% if the retailer takes all the goods. Determine the cost of each suit.

To find the cost of each suit, multiply the cost per piece times the discount and subtract the discount.

$$21.75 \times 15\% = \$3.2625$$

$$\$21.75 - \$3.26 = \$18.49 \quad \text{total per suit}$$

To use your calculator, press

$$\$21.75 \;\boxed{-}\; 15 \;\boxed{\%} \qquad \boxed{18.4875}$$

Don't forget to round. Since we are working with dollars and cents, look to the third place to the right of the decimal. The number is 7, so add 1 to 8. The answer is $18.49 for each suit.

Here's an example for you to try with your calculator.

Determine the cost on the following:

Jogging suits are offered to a leading off-price retailer in June, after the spring season. The suits are offered at the original cost of $48.75 less a 20% seasonal discount if all 212 pieces are purchased.

What is the price for the total purchase? _____

What is the price for an individual suit? _____

Advertising and Promotional Discounts Advertising and promotional discounts are offered to help a merchant advertise or promote merchandise. Often these discounts are based on annual sales volume and the dollars earned are applied directly to an ad, catalog, and/or promotional campaign agreed upon by the vendor and the retailer.

EXAMPLE

A lingerie manufacturer will contribute a 3% discount to a holiday ad based on the total cost of orders placed for the holiday season. The store placed an order for 700 fleece robes costing $37.75 each. How many dollars will be available for a holiday ad?

To find the advertising allowance, first multiply the quantity times the cost per piece; then multiply the total cost by the advertising percentage allowed.

$$700 \times \$37.75 = \$26,425$$
$$\$26,425 \times 3\% = \$792.75$$

To use your calculator, press

$$700 \; \boxed{\times} \; 37.75 \; \boxed{\times} \; 3 \; \boxed{\%} \qquad \boxed{792.75}$$

The advertising allowance is $792.75.

Here's an example for you to try with your calculator.

Determine the advertising dollars available based on the following:

A leading sporting goods firm offers specialty retailers 3% in advertising dollars on the holiday sales placed. The Sports Locker, a local speciality store, placed orders totaling $785,000.00 with the firm last year.

What will be the total advertising dollars earned, based on the advertising discount offered?_____

List Price

Sometimes manufacturers or vendors prefer to set the retail price. They want the consumer to know their product always sells at a specific price. So, let's take a look at *list price,* or MSRP, and then discuss a few more discount opportunities.

List price is also known as the manufacturer's suggested retail price, or MSRP. Manufacturers will establish the selling price to establish consistency and credibility. If this is the case, you will be given the retail price and then deduct a trade discount to determine a cost that you will pay the vendor.

Trade Discounts A **trade discount** is deducted from the list price for both wholesalers and retailers. This discount is offered because the buyer and seller are in the same business, or same trade.

EXAMPLE

A manufacturer offers an item at a list price of $85.00 less a 25% trade discount. Determine the actual cost.

To find the cost, multiply the list price times the 25% trade discount. Then subtract the discount amount from the cost price.

$$\$85.00 \times 25\% = \$21.25$$
$$\$85.00 - \$21.25 = \$63.75 \text{ cost price}$$

To use your calculator, press

85. ⊟ 25 %   63.75

The cost price is $63.75.

Here's an example for you to try with your calculator.

Determine the cost price on the following:

Flannel boxers are sold at a list price of 6.50 per pair. The manufacturer offers them to leading retail stores with a 45% trade discount.

What is the cost of each pair of boxers? _____

Series Discounts Sometimes two or more discounts are deducted from the price, whether it is the cost or the list price that you are quoted. Each discount is given for a reason that the buyer and seller negotiated, and they are *never* combined. Each discount is deducted individually to determine the final buying price. When the discount amounts are written out, they look like a row of fractions.

EXAMPLE

You purchased fabric for $8.50 per yard with discounts of 20/10. How much do you pay?

First, multiply the cost per yard times 20%. Then subtract the discount amount from the cost price.

$$\$8.50 \times 20\% = \$1.70$$
$$\$8.50 - \$1.70 = \$6.80$$

Second, multiply $6.80, which is the cost determined by deducting 20% of $8.50, times 10%. Then subtract the 10% discount amount from $6.80.

$$\$6.80 \times 10\% = \$0.68$$
$$\$6.80 - \$0.68 = \$6.12 \quad \text{cost price}$$

To use your calculator, press

8.50 ⊟ 20 %   ⊟ 10 %   6.12

The cost price per yard is $6.12.

Here's an example for you to try with your calculator.

Determine the price on the following:

A lamp is sold at the MSRP of $45.00 less 25/10.

What is the selling price? _____

Dating and Payment Terms

Now that you have negotiated the cost price, it is time to move on to paying the bill. When a supplier or retailer is paying a bill, two things must always be considered:

1. How much has to be paid
2. When it has to be paid

Now you get to negotiate the second step.

It is very important for you to know when to pay your bill, because you can lower the final cost prices if you pay attention to due dates and earn extra discounts for prompt payment. Let's now take a look at cash discounts and dating terms.

Cash Discount   A **cash discount** is an incentive to pay your bill promptly. It is an additional percent that is deducted from the net amount of the invoice if the bill is paid within a specified period of time. Here are some important features about cash discounts:

1. This additional discount is noted on the purchase order and invoice where you see the word terms.
2. A cash discount can be given in addition to a quantity, trade, or seasonal discount.
3. Cash discounts are calculated after all discounts on the product have been deducted.
4. Cash discounts are given only if the bill is paid within the specified number of days.
5. If the letter *n* or the word *net* is written next to the terms, it means no discount.
6. Cash discount terms never stand alone, but are combined with specific dating terms.
7. When written out, cash discounts and payment days look like fractions, with each number or letter representing a special aspect of how much the discount amount is and how many days the buyer has to earn that discount.
8. Cash discounts are an important way to earn more profit on the merchandise for sale.

Dating **Dating** is the practice of setting a specific time for paying a bill. There are several types of dating, and they work hand in hand with cash discounts. When you see them all written out on a purchase order, you will think you are reading a row of fractions. Sometimes bills have to be paid all at once; sometimes a mill or manufacturer will allow payment at a future date. The most important thing you have to negotiate is exactly *what day* the invoice has to be paid. Here are some key points to note when calculating payment days.

1. There are two ways to calculate the number of days in a month. Even though the months of the year all have different numbers of days, most businesses do the calculations with 30-day months to keep the payment dates easy to calculate. However, many businesses want to take advantage of all possible days in a payment period and therefore follow a traditional calendar.

The number of days in each month are as follows:

January:	31 days	February:	28 or 29 days	March:	31 days
April:	30 days	May:	31 days	June:	30 days
July:	31 days	August:	31 days	September:	30 days
Ocotober:	31 days	November:	30 days	December:	31 days

2. *N* means no discount, so if *N* or *net* appears next to the word terms on a purchase order, the buyer knows that no additional discounts can be earned.
3. If *net* is not indicated on a purchase order, the buyer can assume that there are 30 net days. That is, if the buyer does not take advantage of paying the bill during the extra cash discount days, the bill has to be paid in full within the specified net days. If the buyer does not pay the bill in the specified net days, then the seller is able to charge interest charges or late fees. Usually there are 30 net days given, unless specified otherwise.

Remember, if you ask suppliers or manufacturers for this type of support when buying the finished products, the manufacturers will need to do the same when they purchase materials to make the products. Negotiating is two-sided and both parties, buyers and sellers, have to be able to make a profit.

Dating Terms The following are some of the types of terms you might use for general transactions. These examples identify how the cash incentives are combined with a payment date.

COD means *collect on delivery.* Often this method is used with new accounts. Payment can be cash, company check, cashier's check, or money order. For example, $400.00 COD means collect $400.00 when the goods are delivered.

DOI *(regular)* is often called ordinary dating. The payment window begins by identifying the date of the invoice and calculating the allowed payment days from that date. These are frequently the terms used for supplies that are regularly reordered, such as shopping bags or tissue paper. For example, 2/10, n/30 means a company will earn a 2% discount if the bill is paid within 10 days of the date of the invoice. If the payment is made between 11 and 30 days, the company must pay the full amount.

ROG stands for *receipt of goods.* With so many products being made overseas today, ROG is a very common dating and payment agreement. When using the ROG dating, you do not calculate the days to pay until the merchandise has been received. With merchandise made overseas it is too difficult to calculate when merchandise will be received because of duties and customs in the worldwide ports. For example, an invoice is dated May 5 and the terms read 8/20, n/30 ROG. The merchandise is received on July 3. To receive the extra 8% cash discount, payment must be made on or before July 23.

Note: When the merchandise is received, the shipping company notifies the seller immediately.

As businesses continue to grow and major chain organizations open large storage and distribution centers for shipping and receiving, merchants often buy products in advance of a selling season. There are two different ways to negotiate payment terms so you can get merchandise before the selling season but still have extended payment terms.

Advanced dating is sometimes called *as of dating, future dating,* or *seasonal dating.* These terms simply mean that a date in the future is secured for starting the time to pay and receive the extra cash discount. Often this is used when purchasing for a new store opening: stores need to be filled with inventory but are not ready for retail sales. For example, an invoice is dated June 15. The terms read 8/10 n/30 as of September 5. Merchandise was shipped on June 15. However the 8% cash discount can be taken until September 15, because there is a 10-day period allowed beginning on September 5 (September 5 + 10 days = September 15).

Extra is an allowance of extra days given beyond the normal dating agreements. Often this is used when large quantities are purchased, such as when filling up warehouses to get ready for a holiday season. For example, an invoice is dated September 15 and the terms read 3/10 n/30–60X. The merchandise was shipped on September 15. A 3% cash discount can still be taken as late as November 25 (September 15 + 10 days + 60 days = November 25).

As the retailing and manufacturing industries grew, manufacturers were sending merchandise into many departments throughout a month. It was virtually impossible for a store to pay the invoices on regular terms, so a system for grouping invoices and making a single payment seemed very practical. As a result, EOM terms were initiated; and they are the most popular terms used in the industry.

EOM means *end of month,* which indicates that the invoice must be paid within a specified number of days from the end of the month. For example, an invoice is dated August 15, and the terms read 8/10 n/30 EOM. The merchandise was received on August 22. To receive the cash discount, payment must be made on September 10.

But when is the end of the month? Some months have 30 days, some have 31, and February has 28 or 29. Companies need to have a common cutoff date. In the fashion industry, businesses have assigned the 25th of each month as the cutoff date for the end of the month. That is, if an invoice is dated from the 1st through the 25th of the month, payment is made from the end of that month. For example, an invoice is dated January 18 and terms are 8/10 EOM. Payment is due February 10.

If an invoice is dated from the 26th through the 31st of the month, payment is made from the end of the *next* month. For example, an invoice is dated January 28 and terms are 8/10 EOM. Payment is due March 10. These are the terms many manufacturers and buyers strive for because they give a business almost a 6-week period before the invoice has to be paid and still allow the business to earn an 8% discount.

Freight

Now, let's take a look at the third component, who pays the freight. After you negotiate the lowest possible price and have decided when to pay, the next step is to discuss who pays the shipping charges. Don't forget, fabric mills and manufacturing centers are located throughout the global market, and moving the products around the world costs money. Manufacturers, designers, and retailers all identify the shipping expenses through FOB terms. Let's take a moment to learn what FOB means.

FOB means *freight on board,* or *free on board.* This designation determines when the ownership of goods takes place and, once that exchange occurs, which party is responsible for shipping charges. General shipping terms are as follows.

Industry **Terms** and *jargon*

FOB Store or Destination The manufacturer owns the merchandise and pays all freight charges until the merchandise reaches the store.

FOB Factory or Shipping Point The buyer pays the freight charges, and the amount of transportation charges is added to the invoice.

FOB (city name) or FOB (50/50 or Some other Split)   With foreign goods the transportation charges are often split, and this must be indicated in the negotiations. Sometimes the manufacturer or mill will pay from their shipping point to a designated city and then the designer or store will pick up the shipping expenses from that point to the final destination. For example, if the terms agreed on are FOB New York, the manufacturer has agreed to pay the shipping charges from their location to New York and then the designer or store must pick up the shipping charges to the next designation. Another way to reduce shipping charges is to agree to split the charges. In that case the terms read FOB 50/50.

FOB Prepaid Store or Destination   The manufacturer or vendor pays the shipping charges when the merchandise is shipped and does not charge the store.

FOB Prepaid Factory or Shipping Point   The manufacturer or vendor pays the shipping charges when the merchandise is shipped and then adds the charges to the invoice.

Now that you have reviewed discounts, payment arrangements and shipping terms, the following problems will help you see how good negotiating can result in some good prices!

■ Problems

You can do these calculations using either a standardized 30-day month or a traditional calendar with actual days per month.

1. Determine the last date for the cash discount to be taken.

Invoice Date	Merchandise Received	Terms	Last Day for Discount: Standard 30 Days	Last Day for Discount: Traditional Calendar
March 6	March 10	8/10 EOM	April 10	April 10
March 26	April 8	2/10 ROG		
Sept. 29	Oct. 5	8/10 EOM		
June 17	June 25	6/10–40X		
July 3	July 28	6/10 as of Sept. 20		
Jan. 20	Jan. 28	3/10 EOM		
Aug. 15	Aug. 25	5/10 as of Oct. 20		
May 30	June 5	1/15 n/30		
Feb. 20	Mar. 2	1/10–60X		
Nov. 21	Nov. 26	4/10–50X		
Oct. 26	Dec. 2	2/10 ROG		

2. Merchandise that had a billed cost of $580.00 was shipped on October 12 and received on October 21. The terms were 8/10 EOM. How much should be remitted if paid on each date?

October 30 _____ November 10_____
Which is better, and why? _____

3. A music store purchased a stereo system, which listed at $1,295.00 with discounts of 30% and 15%. The invoice was dated on January 12 with terms of 6/10. If paid on January 20, how much should be remitted?

4. Merchandise was invoiced on April 29 with a total billed cost of $2,249.00. Terms are traditional retailing terms of 8/10 EOM.

   a. What is the last day the cash discount can be taken?

   b. If the invoice is paid on June 10, how much should be paid?

5. Merchandise invoiced on February 18 has a total list price of $678.85 with discounts of 25% and 15%. Terms are 3/10.

   a. How much should be paid on February 28?

   b. How much should be paid on March 10?

   c. What do you suggest, and why?

6. Determine the last date for the cash discount to be taken.

Invoice	Received	Amount	Terms	Amount of Discount	Amount Remitted	Last Day of Discount
3/5	3/9	$ 500	8/10EOM	_____	_____	_____
8/26	9/6	$1,000	2/10ROG	_____	_____	_____
1/15	2/11	$6,345	8/10	_____	_____	_____
10/28	11/4	$ 480	8/10EOM	_____	_____	_____
6/16	6/24	$ 800	6/10–40X	_____	_____	_____
7/5	7/21	$ 300	4/10 as of 9/1	_____	_____	_____
4/10	4/28	$ 785	COD	_____	_____	_____

7. On merchandise with a list price of $5,250.00, a wholesale buyer is offered trade discounts of 30% and 25%. What dollar amount will the buyer be billed?

8. The invoice is dated March 7, and the delivery is March 12. Terms on the purchase order are 8/10 n/30 EOM. What is the last day the buyer can pay the bill and still receive a discount?

9. Merchandise is sent from the manufacturer with the terms of 2/10, n/30, FOB factory. Who pays the shipping charges?

10. An invoice is dated May 12, and merchandise is delivered to the store on May 25. For each of the following terms, determine the dates, using a 30-day business calendar.

	First day of discount	Last day for discount
3/10 net 30	_____	_____
2/10–30X	_____	_____
3/15, n/60 ROG	_____	_____

11. Determine the net cost on the following.

    a. List price $230.00; offered at 45/15 off

    Net cost _____

    b. List price $850.00; offered to trade less 52%

    Net cost _____

12. An invoice for $2,400 with terms of 2/10–30X ROG is received by a specialty store. The invoice is dated May 16, and goods are received on May 21. How much discount can be taken, and what is the last date for the discount?

    Date _____   Amount _____

13. The billed cost on a shipment of hand-blown stemware is $2,000.00 and the freight charges are $80. The delivered cost is $2,080.00. There is a 3% cash discount on this order.

    What is the cash discount? _____

14. A specialty store in Pittsburgh, Pennsylvania places an order to be delivered by April 10. An invoice for $1,800 was prepared on April 5 with terms of 2/10 net 30, FOB Pittsburgh.

    How much would the specialty store owe if the bill were paid April 19?_____

    How much would be owed in shipping charges by the specialty store?_____

15. A bill is for $4,000.00 with terms of 2/10−60X cash discount.

    How much would be remitted if the invoice were paid within the cash discount period? _____

    How many days does the merchant have to pay the bill?_____

16. A buyer places an order for 10 dozen pen sets. The list price is $60.00 each, and discounts of 25%, 20%, and 5% are given. Terms are 2/10 shipped via truck FOB factory. Shipping charges were $28.60 and were prepaid by the vendor. The pen sets were shipped and invoiced on May 18 and received on May 24, in time for the upcoming Father's Day sale.

    How much should be paid on June 5?_____
    How much should be paid on May 26?_____

17. A buyer purchases children's socks for a back-to-school sale. The order is 50 dozen pairs of sweat socks at $8.50 per dozen and 26 dozen pairs of jogging socks at $8.50 per dozen. Any purchases over 40 dozen carry a 5% discount. (*Think:* 40 dozen at $8.50 each dozen and the additional dozens at $8.50 less 5%.)

    What is the net cost?_____

18. 185 national team sweatshirts are sold with a list price of $21.99, less a trade discount of $42\frac{1}{2}\%$.

What is the amount of the discount?_____

What is the net cost?_____

19. The store accountant receives an invoice for $1,400.00 at list price, plus a $54 freight bill with trade discounts of 20 and 10. The invoice is paid on time and earns an additional 2% cash discount. What is the net cost?_____

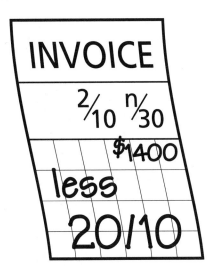

20. Two interior design firms show a similar product at the annual trade show. Company ABC offers the item at $65.00 less 50, 15, and 5. Company CDF offers a similar item at $50 with terms of 25, 20, and 10. Which company offers the best price?

Company ABC price_____

Company CDF price_____

21. What amount should be remitted to the shoe manufacturer for the following merchandise:

16 pairs evening slippers at $24.00 list

36 pairs of water slippers at $12.00 list

24 shoe bags at $10.00 list

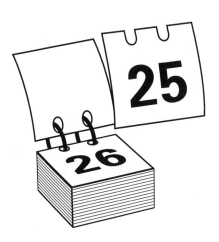

The invoice is dated August 18 and is paid on September 12. Trade discounts are 35 and 15. Terms are 3/10 EOM. FOB store. The vendor has prepaid the transportation charges of $3.78.

How much should be remitted?_____

Explain your decision.

22. The crystal and china buyer purchased the following goods from a leading manufacturer:

Quantity	Item	Individual List Price
4 dozen	Wines	45.00 each
3 dozen	Cordials	36.00 each
$4\frac{1}{2}$ dozen	Goblets	50.00 each

The invoice is dated April 26 and goods are received on May 1. Conditions of sale are as follows: trade discounts of 45/15, terms: 3/10 EOM, FOB store. How much should be remitted and when should it be paid to receive full discounts available?

Date _____   Amount _____

23. An infant buyer completed a foreign buying trip and bought several different gift items. The following was one of the orders that the buyer placed.

Quantity	Item	Individual Cost Price
14 dozen	Silver frames	4.50 each
16 dozen	Silver rattles	65.00 per dozen
20 dozen	Spoon sets	108 per dozen

On orders over $3,000, the manufacturer offers a 5% quantity discount on the entire order. Shipping is FOB NYC. Payment terms are 8/10 ROG. The invoice is dated June 2, and the merchandise is received September 12. How much has to be paid and by what date to receive the full discount?

Date _____   Amount _____

Discuss how the shipping charges will be paid.

24. The linen buyer returned from the Atlanta market and had to decide on a vendor from whom to purchase merchandise for the store's annual August white sale. Based on last year's historical

data, more than $10,000 worth of goods (at cost) were purchased for this major sale. Review the following information and decide what would be the better buy. Discuss your decision.

Linen Company A		Linen Company B	
Item	**Cost**	**Item**	**Cost**
Sheets	$ 5.00	Sheets	$ 5.25
Sheet sets	$12.00	Sheet sets	$12.25
Blankets	$10.00	Blankets	$ 9.00
Terms 2/10 n/30, FOB factory		Terms 8/10 EOM, FOB store	
Orders over $5,000.00, 4% discount		No quantity discounts	

25. The sporting goods buyer purchased:

6 dozen baseball jerseys at $20 each

$12\frac{1}{2}$ dozen baseball pants at $12.50 each

8 dozen baseball caps at $5 each

The invoice is dated May 14 and is paid on June 10. Terms and conditions of sale are 8/10 EOM, and there are FOB factory prepaid transportation charges of $83.76. How much should be remitted? Discuss how the shipping charges will be paid.

Pricing and Repricing Merchandise

four

Individual and Initial Markup

Once merchandise has been purchased, buyers have to price products so customers will want to buy them. There are many factors to consider when pricing, but most successful merchandisers agree it is best to keep it simple and remember the following key points:

1. What is the cost of the merchandise?
   (This is the wholesale or cost price.)
2. What are the operating expenses and how much profit is desired?
   (This is the markup.)
3. Most important, what will the customer be willing to pay?
   (This is the future selling price, either a cost price to a business or a retail price to the consumer.)

You must consider all these factors. Once you have, there are some very basic markup formulas that you can use to help you work within the financial goals of the department and make effective pricing decisions. These formulas involve comparing wholes and parts, applying the same steps you followed with the *Helpful Hints* you have already learned.

Principles of Markup

Industry **Terms** and *jargon*

Cost Price   The price at which the manufacturer sells the merchandise to the retailer. Sometimes this is called the *wholesale price* or, simply, the *cost*.

Retail Price The price the consumer pays.

Markup Also called markon or margin, the amount of money added to the cost price to determine the selling price. Markup has to be adequate to cover expenses **and** make a profit.

Initial Markup The first markup put on merchandise.

Individual Markup The markup for a single item.

Cost (wholesale) Price + Markup = Retail Selling Price The relationship between cost price, markup, and retail price. A buyer typically works with the wholesale cost when purchasing products.

Operating Expenses Expenses incurred by being in business. There are two types, fixed (indirect) and variable (direct).

Fixed Expenses Expenses that cannot be changed. Examples include rent, corporate salaries, and insurance expenses, over which a buyer or department manager has no control. These expenses continue even if the department or area does not exist. Every department in the store contributes to pay these expenses.

Variable Expenses Expenses that affect the operations of a department or area. For example, sales associates' salaries, supplies, and advertising are expenses controlled by the buyer or department manager.

CGS Cost of goods sold, the value of labor and materials needed to make a product. This term is important for buyers in large private label companies who design and produce and then sell a product. Buyers must know the specific value of a product before any markup dollars are added. This is also an important term in calculating profit and loss.

Cost of Goods Sold + Markup = Cost (wholesale) Price The relations between CGS, markup, and cost price. Once a buyer has developed a product and knows the value, markup is added on to reach a cost (wholesale) price, which then serves as the foundation for determining the retail selling price.

Keystone The situation in which markup and cost price are each 50% of the retail price.

California Keystone 52% markup (50% plus 2% extra to help cover shipping costs).

Vendor Manufacturer, wholesaler, supplier, or designer—the person and/or company from which merchandise is bought.

NRF National Retail Federation, originally called the National Retail Dry Goods and then the National Retail Merchants Association. This is the largest trade association in the world. Merchandisers nationwide contribute information, advice, and statistical data shared by members in meetings, reports, and as consultants.

FOR Financial Operating Results which are prepared by the NRF. These results serve as a guide for business owners and merchandisers in measuring performance in multiple areas, such as sales, markdowns, markup, stock-to-sales ratios, and advertising and promotion costs.

Markup is one of the most important concepts in merchandising. It is important that you understand all the elements of this tool, but remember that determining markup is only part of finding the best price.

After studying Chapter 2, you know that all totals are made of parts, and pricing is no exception. Rather than using the terms specific parts and totals, you are now going to use a new vocabulary, since you are dealing with costs, markups, and selling prices. Consider the very basic T-chart grids shown here. By giving everything in both dollars and percents, you can review actual dollars and compare them to your goals. A *T-chart grid* helps you visualize markup. You'll find them very easy to use.

Consider this example.

	Dollars	**Percents**
Cost/wholesale	$24.50	100%
Markup	$12.25	50%
Cost of goods sold	$12.25	50%

The chart states that $24.50 is equal to 100%. That is, when the store buys the merchandise from the manufacturer, the store pays the manufacturer $24.50. Of that $24.50, the CGS is $12.25, or 50% of the cost selling price, and the markup is $12.25, or 50% of the cost selling price.

Look how the *Helpful Hints* are used to calculate and check the work.

Helpful Hint 1
Total amount × percent = specific amount $24.50 × 50% = $12.25

Helpful Hint 2
Specific amount ÷ total amount = percent $12.25 ÷ $24.50 = 50%

Helpful Hint 3
Specific amount ÷ percent = total amount $12.25 ÷ 50% = $24.50

You have now established for how much the designer or manufacturer is selling the product to the store, so let's see for how much the store sells it to the customer.

	Dollars	Percents
Retail	$49.00	100%
Markup	$24.50	50%
Cost/Wholesale	$24.50	50%

The chart shows that the cost is 50% of the retail price and that the markup is also 50% of the retail price. Or, you could say that you have earned $24.50 toward covering expenses and making a profit.

Again, look how the *Helpful Hints* are used to calculate and check the work.

Helpful Hint 1
Total amount × percent = specific amount $49.00 × 50% = $24.50

Helpful Hint 2
Specific amount ÷ total amount = percent $24.50 ÷ $49.00 = 50%

Helpful Hint 3
Specific amount ÷ percent = total amount $24.50 ÷ 50% = $49.00

But, you know that numbers don't come out this even all the time, and it would be impractical to think everything works at a 50% markup. So, let's take a look at the *Helpful Hints* and adapt the ideas of cost, markup, and retail so you can use them in problem solving.

Formulas and T-Chart Grids

Determining markup can be simple with the use of basic formulas. It is also important to use a T-chart grid to

Fill in everything you know,

Solve the obvious, and

Solve the problem.

By using a T-chart grid you can visualize the pricing structure. The following formulas are basic to the concept of markup.

Selling price   = 100% (Once something is sold, the exchange is final and the value is always 100%.)

Cost + markup   = retail (dollars or percents)

Retail – cost   = retail markup (dollars or percents)

Retail – retail markup = cost (dollars or percents)

The following abbreviations are helpful:

R = retail   **MU = markup**   **C = cost**
(total amount)   (specific amount)   (specific amount)

The *Helpful Hints* can be written in terms of retail and cost:

Retail\$ × markup% = markup\$ or   Retail\$ × cost% = cost\$
Markup\$ ÷ retail\$ = markup%   or   Cost\$ ÷ retail\$ = cost%
Markup\$ ÷ markup % = retail\$ or   Cost\$ ÷ cost% = retail\$

A T-chart grid compares the dollar values and percentages:

	Dollars	**Percents**
R =		100%
MU =		
C =		

Finding the Specific Dollar Amounts

Sometimes you will know the retail and the markup or cost percent, but you won't know the markup or cost dollars. Markup and cost are the specific parts of retail:

Retail\$ × markup% = markup\$ or   Retail\$ × cost% = cost\$

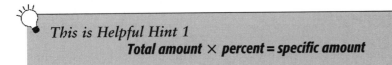

This is Helpful Hint 1
     Total amount × percent = specific amount

E X A M P L E

You purchase merchandise that retails at $20.00. You know retail markup is 60%. What are the retail markup dollars, the cost dollars, and the cost percent?

	Dollars	Percents
R	$20.00	100%
MU		60%
C		

	Dollars	Percents
R	$20.00	100%
MU	$12.00	60%
C	$8.00	40%

1. You know the retail and the markup percent.
2. What is the retail markup in dollars?
3. Calculate the cost dollars and cost percent.

1. Given 100% and 60%
2. $20.00 × 60% = $12.00
3. $20.00 − $12.00 = $8.00
   100% − 60% = 40%

E X A M P L E

You purchase a product to retail for $18.00. You know the cost is 45% of the selling price. What is the cost price, and how much retail markup do you earn in dollars and as a percent?

	Dollars	Percents
R	$18.00	100%
MU		
C		45%

	Dollars	Percents
R	$18.00	100%
MU	$9.90	55%
C	$8.10	45%

1. You know the retail and the cost percent.
2. What are the cost dollars?
3. Calculate the retail markup dollars and percent.

1. Given $18.00 and 45%
2. $18.00 × 45% = $8.10
3. $18.00 − $8.10 = $9.90
   100% − 45% = 55%

Finding Percentages

You know percentages are the most accurate means of comparison, but sometimes they are not given to you. To find percents, simply divide the specific part you have, either the cost or the markup, by the total amount of retail:

Markup\$ ÷ retail\$ = markup%   or   Cost\$ ÷ retail\$ = cost%

This is Helpful Hint 2:
Specific amount ÷ total amount = percent

EXAMPLE

You purchase a product that retails at \$50.00. You know the retail markup is \$20.00. What are the cost dollars and the cost percent?

	Dollars	**Percents**
R	\$50.00	100%
MU	\$20.00	
C		

	Dollars	**Percents**
R	\$50.00	100%
MU	\$20.00	40%
C	\$30.00	60%

1. You know the retail dollars and the markup dollars.
2. What is the retail markup percent?
3. Calculate the cost dollars and cost percent.

1. Given \$50.00 and \$20.00
2. \$20.00 ÷ 50.00 = 40%
3. \$50.00 – \$20.00 = \$30.00
   100% – 40% = 60%

EXAMPLE

You purchase a product to retail for \$80.00. You know the cost is \$42.00. What is the cost percentage, and what is the retail markup in dollars and as a percent?

	Dollars	**Percents**
R	\$80.00	100%
MU		
C	\$42.00	

	Dollars	**Percents**
R	\$80.00	100%
MU	\$38.00	47.5%
C	\$42.00	52.5%

1. You know the retail and the cost dollars.
2. What is the cost percent?
3. Calculate the retail markup dollars and percentage.

1. Given \$80.00 and \$42.00
2. \$42.00 ÷ 80.00 = 52.5%
3. \$80.00 – \$42.00 = \$38.00
   100% – 52.5% = 47.5%

Finding Total Dollar Amounts

Sometimes you know the specific retail markup or cost dollars and also know the percent equivalent; however, you don't know the total retail amount. Recall the following:

Markup\$ ÷ markup% = retail\$ or   Cost\$ ÷ cost% = retail\$

This is Helpful Hint 3
    Specific amount ÷ percent = total amount.

E X A M P L E

You purchase a product on which retail markup is \$22.00, or 44%. What is the selling price? What are the cost dollars and the cost percent?

	Dollars	**Percents**
R=		100%
MU=	\$22.00	44%
C=		

	Dollars	**Percents**
R=	\$50.00	100%
MU=	\$22.00	44%
C=	\$28.00	56%

1. You know the retail markup in dollars and as a percent.
2. What is the retail?
3. Calculate the cost dollars and cost percent.

1. Given \$22.00 and 44%
2. \$22.00 ÷ 44% = \$50.00
3. \$50.00 − \$22.00 = \$28.00
   100% − 44% = 56%

EXAMPLE

You purchase a product at a cost of $16.00, which is 40% of the selling price. What is the selling price and what is the retail markup in dollars and as a percent?

	Dollars	Percents
R		100%
MU		
C	$16.00	40%

	Dollars	Percents
R	$40.00	100%
MU	$24.00	60%
C	$16.00	40%

1. You know the cost dollars and percent.
2. What is the retail?
3. Calculate the retail markup dollars and percent.

1. Given $16.00 and 40%
2. $16.00 ÷ 40% = $40.00
3. $40.00 – $16.00 = $24.00
   100% – 60% = 40%

■ Problems

When working in the industry you are not always given all the facts and figures, and you are left to figure out the missing pieces on your own. When trying to determine prices, markups, and percentages, draw a basic T-chart grid, fill in what you know, solve for the obvious, and use the formulas to find the remaining information.

1. A buyer was trying to determine values on items for sale in the marketplace. In some instances the buyer didn't know the cost, the retail, or the retail markup percentage. Find the following values by using the following T-chart grids and the basic formulas. (Remember, R = retail, MU = markup, and C = cost.) Part (a) has been completed for you.

		Retail Dollars	Cost Dollars	Retail Markup Dollars	Retail Markup Percentage
a.		$ 68.00	$ 23.00	_____	_____
b.		$465.50	$169.99	_____	_____
c.		$440.00	_____	_____	75%
d.		_____	$120.00	_____	48%
e.		$ 14.50	$ 7.83	_____	_____
f.		_____	$100.00	_____	44%
g.		_____(each)	$ 30.00/doz.	_____	54%
h.		_____(each)	$ 18.00/doz.	_____	50%

a.

	Dollars	Percents
R =	$68.00	100%
MU =	$45.00	66.18%
C =	$23.00	33.82%

1. $68.00 − $23.00 = $45.00
2. $45.00 ÷ $68.00 = 66.18%
3. 100% − 66.18% = 33.82%

b.

	Dollars	Percents
R =		100%
MU =		
C =		

c.

	Dollars	Percents
R =		100%
MU =		
C =		

d.

	Dollars	Percents
R =		100%
MU =		
C =		

e.

	Dollars	Percents
R =		100%
MU =		
C =		

f.

	Dollars	Percents
R =		100%
MU =		
C =		

g.

	Dollars	Percents
R =		100%
MU =		
C =		

h.

	Dollars	Percents
R =		100%
MU =		
C =		

Use a T-chart grid to help you solve the following problems.
Remember:

- List everything you are given.
- Identify what you want to know.
- Complete any obvious calculations.
- Look at the formulas and determine what you need to find the remaining values.
- Complete the solution.
- Look at all the numbers. Is the answer logical?
- Check your work.

2. Bookcases are purchased by a retail merchant for $215.00 and are to be placed in the store with a retail selling price of $385.00. Find the markup percentage on retail that she expects to obtain on the bookcases.

	Dollars	Percents
R		100%
MU		
C		

3. A buyer has established a markup on retail of 33.3% for a line of novelty items. The buyer plans to price them at $6.00. Determine the price he will have to pay to maintain his plan.

	Dollars	Percents
R		100%
MU		
C		

4. As an assistant buyer you have been asked to determine the cost and retail price on an item that your department has purchased. The item will carry a $65.00, or 47%, markup. Find the retail price rounded to the nearest dollar.

	Dollars	**Percents**
R =		100%
MU		
C =		

5. A buyer of women's jewelry purchased a line of tennis bracelets to retail at $150.00. The manufacturer has offered these bracelets to the retailer at a cost of $60.00 each. Find the markup percent that the merchant will use.

	Dollars	**Percents**
R =		100%
MU		
C =		

6. A sporting goods firm purchases knit tops. The markup will be 54.5%, or $5.45. What is the retail? What is the cost?

	Dollars	**Percents**
R =		100%
MU		
C =		

7. The cost of a shirt is $12.75. The buyer wants a 49% markup. What is the retail?

	Dollars	**Percents**
R		100%
MU		
C		

8. Determine the markup both in dollars and as a percent for a cedar bedding chest that wholesaled at $625.00 and retailed at $1,400.00.

	Dollars	**Percents**
R		100%
MU		
C		

9. Determine the markup for glasses costing $132.00 per dozen and retailing at $22.00 each.

	Dollars	**Percents**
R		100%
MU		
C		

10. What is the cost of a scarf that retails for $18.00 and has been marked up 40.5%?

	Dollars	**Percents**
$\underline{\underline{R}}$		$\underline{\underline{100\%}}$
$\underline{\underline{MU}}$		
$\underline{\underline{C}}$		

Solve the following problems without using a T-chart grid.

11. If the markup is $23.00 and the retail is $51.50, what is the cost in dollars and as a percent?

12. Roller blades cost $80 and retail for $175. What is the retail markup percent?

13. What is the cost percent of glow necklaces that cost $9.00 per dozen and sell for $2.00 each?

14. Retail is $75.00, markup is $45. MU% is _____ .

    Cost is $30. Retail is $65. MU$ = _____ , MU% = _____ .

    Cost is $12. MU is $10. Retail = _____ , MU% = _____ .

15. Leather tote bags cost the department buyer $90.00 each. In order to obtain a 40% markup, what is the retail price of the totes?

16. A golf bag cost $124.75. The store wants a 36% markup. What is the retail price, rounded to the nearest dollar?

17. A book store retails paperback novels for $3.95. These novels carry a 58% markup. What is the cost price?

18. A designer sells belts at $48.00 per dozen. The belts are retailed at $10.00 each. What is the markup earned in dollars and as a percent?

19. If the cost of an item is $84 and that is a 60% cost figure, what is the retail selling price?

20. Coffee mugs cost $6.00 per dozen and sell for $1.50 each. What is the markup on the coffee mugs in both dollars and as a percent?

In the following, you must first extend the totals, just as though you were completing a purchase order, because you need to calculate the markup value of the total order, not just of the individual pieces. This means:

Quantity × individual cost   = total cost
Quantity × individual retail = total retail

By determining the markup dollars and percent on a *total purchase,* a buyer can tell how close the values of the items are to the departmental markup goal.

21. Determine the markup dollars and percent on the following order. The problem has been started for you. (Don't forget the dozen calculations.)

12 dozen pens   Cost: $12.60 per dozen   Retail: $1.89 ea.
27 dozen glitter pens   Cost: $13.20 per dozen   Retail: $1.99 ea.
11 dozen football pens   Cost: $ 9.00 per dozen   Retail: $1.29 ea.

Remember dozens:
12 × $12.60 = $151.20 12 × 12 = 144 and 144 × $1.89 = $272.16

Units	Cost Each	Cost Total	Retail Each	Retail Total
12 doz./144	$12.60 doz.	$151.20	$1.89	$272.16
27 doz./324	$13.20 doz.	_____	$1.99	_____
11 doz./132	$9.00 doz.	_____	$1.29	_____
Total 50 doz./600		_____		_____

Put the cost and retail totals in the T-chart grids and find the overall markup dollars and percent.

	Dollars	Percents
R=		100%
MU=		
C=		

22. Determine the markup dollars and percent on the following order.

Quantity	Cost Dollars	Retail Dollars
120 pr leotards	$48.00/doz.	$8.50 each
120 pr exercise tights	$ 3.50 each	$6.50 each

	Units	Cost Each	Cost Total	Retail Each	Retail Total
	_____	_____	_____	_____	_____
	_____	_____	_____	_____	_____
Total	_____		_____		_____

Put the cost and retail totals in the T-chart and find the overall markup dollars and percent.

	Dollars	Percents
R=		100%
MU=		
C=		

23. Determine the markup dollars and percent on the following order.

Quantity	Cost Dollars	Retail Dollars
30 coats	$54.00	$108.00
24 sweaters	$28.00	$ 60.00
18 scarves	$16.00	$ 32.00

	Units	Cost Each	Cost Total	Retail Each	Retail Total
	_____	_____	_____	_____	_____
	_____	_____	_____	_____	_____
	_____	_____	_____	_____	_____
Total	_____		_____		_____

Put the cost and retail totals in the T-chart grid and find the overall markup dollars and percent.

	Dollars	**Percents**
R =		100%
MU =		
C =		

24. Determine the markup dollars and percent on the following order.

Quantity	Cost Dollars	Retail Dollars
$2\frac{1}{2}$ dozen	$132.00/ doz.	$20.00 each
$1\frac{1}{6}$ dozen	$ 15.00 each	$28.00 each
$3\frac{1}{3}$ dozen	$ 12.00 each	$25.00 each

	Units	Cost Each	Cost Total	Retail Each	Retail Total
	_____	_____	_____	_____	_____
	_____	_____	_____	_____	_____
	_____	_____	_____	_____	_____
Total	_____		_____		_____

Put the cost and retail totals in a T-chart grid and find the overall markup dollars and percent.

	Dollars	Percents
R =		100%
MU =		
C =		

Solve the following problems. First you might have to use T-chart grids to help you find all the parts that you will need to extend the order. Once you have all the parts to extend the order, calculate the overall markup dollars and percent for the order.

25. The small-leather buyer purchased the following items:

200 small leather attache bags at a cost of $36.00 that will retail at $80.00.

150 men's travel kits at $4.00 each; 52.3% markup.

350 saddle style backpacks to retail at $120.00; 56.2% markup.

(*Remember*: Combine everything you know so far. List everything you know. Complete the obvious and then solve the problem: What is the overall markup on this order? Use your *Helpful Hints* and *Markup Hints* to help you find the parts that are missing. Use the T-chart grids to determine individual values, and then complete the total chart.)

	Dollars	Percents
R		100%
MU		52.3%
C	$4.00	

	Dollars	Percents
R	$120.00	100%
MU		56.2%
C		

Units	Cost Each	Cost Total	Retail Each	Retail Total
200	_____	_____	_____	_____
150	_____	_____	_____	_____
350	_____	_____	_____	_____
Total _____		_____		_____

	Dollars	Percents	Work Area
R		100%	
MU			
C			

26. The junior-sportswear buyer purchased 65 T-shirt dresses, getting a 58.5% markup with $36.00 in markup dollars on each. She also bought 36 cardigan jacket tops for $42.00 that she plans to retail at $80.00 each. Determine the overall markup on this order.

(*Remember:* Combine everything you know so far. List everything you know. Complete the obvious and then solve the problem: What is the overall markup on this order? Use **your *Helpful Hints* and *Markup Hints* to help you find the parts that are missing. Use the T-chart grids to determine individual values, and then complete the total chart.**)

Dollars   Percents

R
=   100%

MU
=

C
=

Use this T-chart grid to help you determine the cost and retail of the T-shirt dresses.

Units	Cost Each	Cost Total	Retail Each	Retail Total
_____	_____	_____	_____	_____
_____	_____	_____	_____	_____

Total _____   _____   _____

Dollars   Percents   **Work Area**

R
=   100%

MU
=

C
=

What retail price do you think the buyer should charge for the T-shirt dresses?

27. The jewelry buyer bought the following:

180 chains at $108.00 per dozen that will retail with a 52.5% markup.

200 pairs earrings with a markup of $15.00 (53.57%).

280 rings to sell at $36.00 retail with a 55% markup.

What is the markup percent on the overall order?

	Dollars	Percents
R		100%
MU		
C		

	Dollars	Percents
R		100%
MU		
C		

	Dollars	Percents
R		100%
MU		
C		

Use these t-charts to help you determine the individual cost and retail prices before calculating the overall order.

	Units	Cost Each	Cost Total	Retail Each	Retail Total
	_____	_____	_____	_____	_____
	_____	_____	_____	_____	_____
	_____	_____	_____	_____	_____
Total	_____		_____		_____

	Dollars	Percents	Work Area
R		100%	
MU			
C			

Use this T-chart grid for the totals.

28. Use the purchase order (the form covered in Chapter 3) on the following page.

a. Fill in all the information available.
   Date: August 5, 19__
   Purchase order no. 514
   Store name: Shephard's Specialty Store
                            908 E. Olman Avenue
                            Tampa, FL 39999
                            Phone: (407) 999-6789
                            Fax: (407) 999-7890

   The main store address is the same as the shipping address.
   Cruise Wear, department 480

   Deliver December 25, not before. Cancel if not received by
   January 5.
   FOB factory prepaid, via UPS.
   Terms 8/10 EOM.

   Vendor name: San Marco Fashions
                            500 Delmar
                            Miami Shores, FL 33333
                            Vendor no. 29250
                            Phone: (305) 777-4411
                            Fax: (305) 777-4422

b. Extend the order. The buyer purchased:

1 dozen terry cover-ups, one size fits all; cost $12.50 each; retail $25.00 each.
2 dozen swim caps, one size fits all; cost $24.00 per dozen; retail $ 5.00 each.

c. Determine the individual retail markup percent on the items ordered.

d. Calculate the overall retail markup percent on the order by using the total cost dollars and total retail dollars.

e. The planned departmental markup is 48%. Discuss how the overall markup compares to the planned markup.

PURCHASE ORDER

P.O. no. _____

Store Name

Address _____

Phone _____ Fax _____

Vendor name _____

Vendor address _____

Vendor city, state, ZIP _____

Vendor phone, fax numbers _____

Vendor no. _____

Main store and branch store addresses

Ship to _____

Ship to _____

Ship to _____

Order date _____

Department name _____

Department no. _____

Shipping date _____

Cancel if not received by _____

FOB _____

Ship via _____

Terms _____

Order MU %

Planned Departmental MU %

Original and 3 copies
1. Original to vendor/manufacturer
2. Copy to store accounting department
3. Copy to store receiving department
4. Copy to buyer's records

<u>Note</u>: The unit retail, total retail, MU $, and MU % are hidden from the manufacturer.

The original shows only the cost values, because that is what the store is paying.

Qty	Style	Description	Size	Color	Unit Cost	Unit Retail	Total Cost	Total Retail	Total MU $	MU % Each
Totals							Total Cost	Total Retail	Total MU $	Overall MU %

29. Using the purchase order (the form covered in Chapter 3) on the following page.

    a. Fill in all the information available.

        Date: July 15, 19__
        Purchase order no. 318
        Store name: The Kid's Factory
                      102 Pawnee Place
                      Wichita, KS 67255
                      Phone: (316) 772-2332
                      Fax: (316) 772-2333

        Main store address is the same as the shipping address.
        Hand-held electronics department, no. 990

        Deliver October 15. Cancel if not received by October 30.
        FOB NYC, via ship/truck.
        Terms 8/10 ROG-30X

        Vendor name: Madison Toy Company
                       P.O. Box 65
                       New York, NY 10002
                       Vendor no. 982
                       Phone: (212) 486-7878
                       Fax: (212) 486-7858

    b. Extend the order. The buyer purchased:

12 dozen hand-held video games; cost $14.75 each; retail $25.00 each
 6 dozen game cartridges; cost $35.75 each; retail $65.00 each
 6 dozen game cartridges; cost $25.75 each; retail $55.00 each
10 dozen set of batteries; cost $ 1.50 each; retail $ 2.75 each

    c. Determine the individual retail markup percent on the items ordered.

    d. Calculate the overall retail markup percent on the order by using the total cost dollars and total retail dollars.

    e. The planned departmental markup is 45%. Discuss how the overall markup compares to the planned markup.

PURCHASE ORDER

P.O. no _____

Store Name

Address

Phone _____ **Fax** _____

Vendor name _____

Vendor address _____

Vendor city, state, ZIP _____

Vendor phone, fax numbers _____

Vendor no. _____

Main store and branch store addresses

Ship to _____

Ship to _____

Ship to _____

Order date _____

Department name _____

Department no. _____

Shipping date _____

Cancel if not received by _____

FOB _____

Ship via _____

Terms _____

Order MU %	Qty	Style	Description	Size	Color	Unit Cost	Unit Retail	Total Cost	Total Retail	Total MU $	MU % Each
Totals								Total Cost	Total Retail	Total MU $	Overall MU %

Planned Departmental MU %

Original and 3 copies
1. Original to vendor/manufacturer
2. Copy to store accounting department
3. Copy to store receiving department
4. Copy to buyer's records

<u>Note</u>: The unit retail, total retail, MU $, and MU % are hidden from the manufacturer.

The original shows only the cost values, because that is what the store is paying.

30. Using the purchase order (the form covered in Chapter 3) on the following page.

    a. Fill in all the information available.
        Date: January 10, 19__
        Purchase order no. 318
        Store name: The Clothes Trunk
                     1345 Market Street
                     Boardman, CT 12333
                     Phone: (330) 726-9723
                     Fax: (330) 726-9733

       Main store address is the same as the shipping address.
       Men's Activewear Department no. 58

       Deliver May 15. Cancel if not received by May 20.
       FOB store, via truck.
       Terms 2/10 n/30.

       Vendor name: Tom's Closet
                     777 5th Avenue
                     New York, NY 10022
                     Vendor no. 58675
                     Phone: (212) 654-3225
                     Fax: (212) 654-3335

    b. Extend the order. The buyer purchased:

12 men's Henley style shirts in ecru; 3S, 3M, 3L, 3XL; cost $24.74 each; retail $45.00
 6 pair drawstring ecru linen pants; 1S, 2M, 2L, 1XL; cost $25.75 each; retail $50.00
 6 windbreaker-style ecru linen jackets; 2M, 2L, 2XL; cost $32.75 each; retail $65.00

    c. Determine the individual retail markup percent on the items ordered.

    d. Calculate the overall retail markup percent on the order by using the total cost dollars and total retail dollars.

    e. The planned departmental markup is $47.5%. Discuss how the overall markup compares to the planned markup.

PURCHASE ORDER

P.O. no _____

Store Name

Address _____

Phone _____ Fax _____

Vendor name

Vendor address

Vendor city, state, ZIP

Vendor phone, fax numbers

Vendor no.

Main store and branch store addresses

Order date _____

Department name _____

Department no. _____

Shipping date _____

Cancel if not received by _____

Ship to _____ FOB _____

Ship to _____ Ship via _____

Ship to _____ Terms _____

Qty	Style	Description	Size	Color	Unit Cost	Unit Retail	Total Cost	Total Retail	Total MU $	MU % Each
Totals						Total Cost	Total Retail	Total MU $	Overall MU %	

Order MU %

Planned Departmental MU %

Original and 3 copies
1. Original to vendor/manufacturer
2. Copy to store accounting department
3. Copy to store receiving department
4. Copy to buyer's records

<u>Note</u>: The unit retail, total retail, MU $, and MU % are hidden from the manufacturer.

The original shows only the cost values, because that is what the store is paying.

five

Average, Cumulative, and Maintained Markup

So far we have focused on individual pieces and values and then putting all the parts together. However, a buyer rarely looks at only one item to purchase at a time. Usually, a merchant will look at groups or lines of merchandise and buy an assortment of items with a variety of cost and retail prices. The merchandise purchased also has to mix in well with what is on the selling floor. The goal is always the same: to buy the right merchandise for the right price and place it in the stores at the right time to achieve maximum sales results, while still meeting the customer's needs.

Chapter 5 is about blending merchandise—combining quantities of items with various costs and retail prices to reach a desired markup known as the **departmental markup.** All departments have markup goals. The departmental markup is the target figure for which a merchandiser strives when placing orders. If the merchant maintains the departmental markup on the purchases, there will be enough markup dollars overall to cover department expenses and still make a profit.

If you are starting a new business or if you want to check industry standards, you can use the financial operating results (FOR) prepared annually by the National Retail Federation (NRF). These reports will identify what a typical departmental markup should be on a certain category of merchandise, by store volume.

Since every item doesn't carry the same markup, buyers will purchase merchandise and blend prices together to meet their goals. The following illustrates three different ways that buyers can approach their pricing decisions.

Average Markup

The most common markup planning is called **average markup.** With average markup, the buyer strives to successfully blend quantities of merchandise together at a variety of costs and retail to reach an overall markup that is within the departmental goals. A merchant compares the average markup achieved on purchases to the departmental markup to see if the items purchased will provide enough markup to meet the goal. Buyers can easily see how the average markup on a group works out on a single purchase order, just as you did when calculating the markup of an order in the problems at the end of Chapter 4. To see the overall picture, merchants rely on a tool called a **buying plan** before making purchases. A buying plan allows merchandisers the opportunity to

Review the department goals.

Identify the merchandise that is on hand.

Review what is on order.

Determine the stock that needs to be purchased.

Plan how the merchandise should be priced.

EXAMPLE

Let's consider a sportswear group a buyer saw in the fall line at market and wants to purchase. The buyer needs to know once all the prices have been averaged if this merchandise will meet the departmental goals of a 48% markup. She wants to buy

- 12 blazers with a cost of $55.75 each to retail at $100.00 each
- 18 skirts with a cost of $22.75 each to retail at $45.00 each

In addition, the buyer wants to purchase 36 coordinating turtlenecks that cost $14.75, but she is not sure of the retail price.

First, as the buyer does, fill in everything you know about the merchandise and put that information in the buying plan. In this case the buyer wants 12 blazers that cost $55.75 each and retail at $100.00 each. Rather than work with individual values and pieces, you will work with the totals, just as you did with the problems in Chapter 4.

$$12 \times \$55.75 = \underline{\$669} \quad \text{and} \quad 12 \times \$100.00 = \underline{\$1,200}$$

Then she wants to buy 18 skirts at $22.75 each with a retail price of $45.00.

$$18 \times \$22.75 = \underline{\$409.50} \quad \text{and} \quad 18 \times \$45 = \underline{\$810}$$

Fill these values in a buying plan.

100%

	Units	Cost Each	C$ Total	Retail Each	R$ Total	MU%	C%
On order	12	$55.75	$669.00	$100.00	$1,200.00	—	—
On order	18	$22.75	$409.50	$45.00	$810.00	—	—
On order	—	—	—	—	—	—	—
Dept. goals and/or plans	—	—	—	—	—	—	—

Now, put in the information you have on the turtlenecks, and complete any other obvious calculations. And don't forget, your retail markup for the department is 48%. Thus, the value of the cost percent on this order is 52%. After you have entered the quantity and cost of the turtlenecks, your chart should look like this.

100%

	Units	Cost Each	C$ Total	Retail Each	R$ Total	MU%	C%
On order	12	$55.75	$669.00	$100.00	$1,200.00	—	—
On order	18	$22.75	$409.50	$45.00	$810.00	—	—
On order	36	$14.75	$531.00	—	—	—	—
Dept. goals and/or plans	66	—	$1,609.50	—	—	48%	52%

Now it is time to follow the step-by-step approach you used in Chapter 4 to solve the problem. You want to find the retail price needed on the turtlenecks to achieve an average markup of 48% and be in line with the departmental goal.

The total cost is $1,609.50   .

The cost% is   52%   .

Total cost$ ÷ C% = $3,095.19   total retail$.

Total retail $ − retail $ spent on the jackets and skirts = $3,095.19 − $1,200 − $810 = $1,085.19.

Remaining retail $ open allocated for the turtlenecks = $1,085.19.

  $1,085.19   retail $ open ÷ 36 turtlenecks = $30.14   retail price each.

The buyer knows if she prices the turtlenecks at ___$30.14___ , the overall average on the group of merchandise will meet the departmental goals, even though each item has a very different individual markup. ■

Talk Out

Of course, $30.14 is not a realistic price, and the buyer now has to make a decision. Does she price the turtleneck $30.00, a bit shy of the goal, or does she raise the price to $31.00? What do you think the buyer will do? Ask some tough questions, starting with:

Do you think the customer is willing to pay the price for this merchandise?

Could you raise the price of one of the other pieces?

This step-by-step process isn't limited only to new line purchases. Quite often you will need to adapt this process to find out how much you should buy, based on purchases already made or company goals that have been set. Let's take a look at a typical situation that you, as a buyer, may encounter.

EXAMPLE

A buyer of small-leather goods plans a special in-store promotion to sell 60 dozen calendar planners for $20.00 each. He already has on hand 32 dozen that cost $9.75 each. He still needs to buy another 28 dozen to meet his sales goals. What is the most he can pay for the other 28 dozen calendar planners and still obtain a departmental markup of 47.5%?

Following the same steps as before, list everything you are given in the following chart, identify what you are missing, and complete any obvious calculations.

100%

	Units	Cost Each	C$ Total	Retail Each	R$ Total	MU%	C%
Dept. goals and/or plans	60 doz./720	____	____	$20.00	$14,400	47.5%	52.5%
On hand	32 doz./384	$9.75	$3,744	____	____	____	____
Open to buy	28 doz./336	____	____	____	____	____	____

Following the basic markup formulas you have used throughout this workbook, fill in the remaining values. How much will the buyer need to pay for the 28 dozen calendar planners? Complete only the sections needed. Remember, MU% and C% together must equal 100%.

Your chart should look like this.

	Units	Cost Each	C$ Total	Retail Each	R$ Total	MU%	C%
Dept. goals and/or plans	60 doz./720	⎯⎯	⎯⎯	$20.00	$14,400	47.5%	52.5%
On hand	32 doz./384	$ 9.75	$3,744	⎯⎯	⎯⎯	⎯⎯	⎯⎯
Open to buy	28 doz./336	$11.36	$3,816	⎯⎯	⎯⎯	⎯⎯	⎯⎯

Use your calculator and complete the chart above as you determine each answer.

14400 ☒ 52.5 % ⊟ 3744 ÷ 336 `11.357142`

The price on the remaining 28 dozen calendars needs to be __$11.36__ .

Talk Out

Again, $11.36 is not a realistic price, but what would be the price range the buyer would be looking for in the market?

Will the quality be similar in the products? Discuss what the buyer might do when the merchandise is displayed on the selling floor. ■

Cumulative Markup

It's time to look at another pricing trend. With the industry rapidly opening more outlet stores, such as Ann Taylor and Levi's, along with off-price giants such as Marshalls, T.J. Maxx, and Loehmann's, cumulative markup is very common. **Cumulative markup** involves taking a group or lot of merchandise with a variety of cost values and determining one retail price. Consumers often see similar merchandise with one price, although it appears some merchandise is better than other merchandise. This situation occurs because strong quantity and off-season discount negotiations determine competitive cost prices, and cumulative markup is the retail pricing method used to identify a common retail value.

Cumulative markup is very easy to calculate, because you are always working with one retail selling price. Let's look at the following and see how you group the cost values and look for a common retail price that will achieve the average departmental goal.

EXAMPLE

At the end of the spring market a major manufacturer was offering the remaining inventory of swimsuits from the season's cruise line at the following prices:

16 sarong suits at $22.00 each

32 tank suits at $18.00 each

16 two-piece bikini prints at $14.00 each

27 two-piece multiprints at $15.00 each

All the bathing suits are purchased by the sportswear buyer, and they are to be retailed at the same price. What retail price will result in a 48% markup?

Begin by determining the total cost.

$16 \times \$22.00 =$ ___$352.00___

$32 \times \$18.00 =$ ___$576.00___

$16 \times \$14.00 =$ ___$224.00___

$27 \times \$15.00 =$ ___$405.00___

Total pieces ___91___ Total cost ___$1557.00___

Now, using the T-chart grid, fill in the total cost and the retail markup of 48%.

	Dollars	Percents
R		100%
MU		48%
C	$1,557	

Once you have those figures filled in, complete the calculations to finish the T-chart grid. Your chart should look like this:

	Dollars	Percents
R	$2,994.23	100%
MU	$1,437.23	48%
C	$1,557.00	52%

$1,557 ÷ 52% = $2,994.23

$2,994.23 − $1,557.00 = $1,437.23

100% − 48% = 52%

Once you know the total retail, simply divide the total retail dollars by the total number of pieces to determine the individual selling price.

Total retail dollars ÷ total number of pieces = individual price

$$\$2,994.23 \div 91 = \$32.90$$

___$32.90___ is the retail selling price needed to reach departmental goals.

Talk Out

Based on the time of year that you are purchasing this swimwear, what could be the price ceiling on these goods? If you did mark the goods higher than the minimum price, what would that do for the overall operation? ■

Maintained Markup

When buyers decide if a line of merchandise has sold successfully, they will use the company vendor analysis reports or classifications reports (or both) to identify the maintained markup. Simply put, **maintained markup** is the final markup achieved after all the merchandise has been sold, no matter what the selling price, compared to the original cost of the merchandise. Merchandise may be priced and repriced in order to give the most successful selling results in a company. Notice we say *repriced*—not all merchandise is marked down. Sometimes merchandise is marked up after a promotional sale. Such pricing decisions are all part of the pricing strategy to encourage sales while still looking for a means of meeting the goals initially established by the owners or buyers of achieving profitable markups.

Let's work on a problem so you can see how the results are determined. And, when doing these problems, always refer to the initial markups taken as a starting point for comparison.

EXAMPLE

A buyer purchased 500 pairs of leather gloves at $7.50 per pair and retailed them for $15.00 per pair. 300 pairs were sold at the original retail price. The 200 remaining pairs were marked down to $12.00 each. At this price 120 pairs were sold. The remaining 80 pieces were all sold at a half-off clearance sale for $7.50 per pair. First you should determine the initial markup percent and then compare the results to the maintained markup percent.

Use the following T-chart grids.

List everything you are given.

Identify what you want to know.

Complete any obvious calculations.

Look at the formulas and determine what you need to find the remaining values.

Complete the solution.

Look at all the numbers. Is the answer logical?

Check your work.

Fill in the T-chart grid for initial markup with this information:

500 × $7.50 = <u>$3,750.00</u> and 500 × $15.00 = <u>$7,500.00</u>

$7,500.00 – $3,750.00 = $3,750.00 retail markup dollars

$3,750.00 ÷ $7,500.00 = 50%

Initial Markup

	Dollars	**Percents**
R	$7,500.00	100%
MU	$3,750.00	50%
C	$3,750.00	50%

Once you negotiate the cost value, that price never changes, but the retail price can. Using the following T-chart grid below, fill in the cost price and the total retail value of all the sales results, and determine the maintained markup for this purchase.

300 pairs were sold at $15.00:   300 × $15.00 = <u>$4,500.00</u>

120 pairs were sold at $12.00:   120 × $12.00 = <u>$1,440.00</u>

80 pairs were sold at $ 7.50:   80 × $ 7.50 = <u>$600.00</u>

Adding the retail selling figures gives final retail sales of $6,540.00, which is the retail figure you need to calculate the maintained MU%.

Maintained Markup

	Dollars	**Percents**
R	$6,540.00	100%
MU	$2,790.00	42.66%
C	$3,750.00	57.33%

Use your calculator:

6540 $\boxminus$ 3750 $\boxdiv$ 6540 $\boxed{\%}$ `42.66055`

100 $\boxminus$ 42.66 $\boxdot$ `57.33`

Talk Out

If the departmental goal is 47%, what is your opinion on this purchase? What are some questions you might want to ask about the selling results? Would you need to discuss the climate this past year? What about colors, styles, and quantities? This information is so important, because from the vendor analysis and classification reports we can learn about the consumer's buying habits. The buyer did not necessarily make a "bad" buy, but some adjustments might be needed for next season. ■

Not all repricing involves markdowns. Sometimes the prices will go up instead of down. Simply follow the same steps and then compare the initial markup to the maintained markup and look at the overall success of the purchase.

■ Problems

Average Markup

As you can see from the examples, you will use the same formulas, along with your calculator to help you out, but now you will think ahead a bit, applying strong critical thinking skills to determine the pricing or quantities that you need to reach departmental goals. Remember, combine everything you know so far and follow these steps:

List everything you are given.

Identify what you want to know.

Complete any obvious calculations.

Look at the formulas and determine what you need to find the remaining values.

Complete the solution.

Look at all the numbers. Is the answer logical?

Check your work.

(Note: complete only the sections needed)

1. A lingerie buyer plans to purchase 48 dozen kimonos to retail for $25.00 each. She already placed an order for 26 dozen at a cost of $12.50 each. What is the most she can pay for the remaining kimonos and still obtain a departmental markup of 48%?

100%

	Units	Cost Each	C$ Total	Retail Each	R$ Total	MU%	C%
Dept. goals and/or plans	____	____	____	____	____	____	____
On order	____	____	____	____	____	____	____
Open to buy	____	____	____	____	____	____	____

2. A jewelry buyer intends to have enough merchandise on hand to sell for $14,000 total retail value, with a departmental markup of 49%. After a New York buying trip, he has already purchased 200 pairs of earrings to retail at $32.00 a pair with a 46% markup. How much money does the buyer have to spend at both cost and retail to achieve a departmental markup of 49%?

100%

	Units	Cost Each	C$ Total	Retail Each	R$ Total	MU%	C%
Dept. goals and/or plans	____	____	____	____	____	____	____
On order	____	____	____	____	____	____	____
Open to buy	____	____	____	____	____	____	____

*Sometimes the cost percentage is referred to as the *markup complement*.

3. A stationary buyer was placing orders for novelty boxes of stationery. Departmental markup for stationery is 52%. The buyer has purchased the following:

4 dozen boxes of children's note cards at $10.00 per dozen to retail at $1.75 each

6 dozen silk-screened thank-you note cards at $15.00 per dozen to retail at $3.00 each

She also has purchased 6 dozen plain note cards at $12.00 per dozen. What should be the retail selling price of the plain note cards for the purchase order to achieve an average markup of 52% to meet the departmental goals?

100%

	Units	Cost Each	C$ Total	Retail Each	R$ Total	MU%	C%
On order	____	____	____	____	____	____	____
On order	____	____	____	____	____	____	____
On order	____	____	____	____	____	____	____
Dept. goals and/or plans	____	____	____	____	____	____	____

What price do you think the buyer would set on the plain note cards?

4. A toy buyer plans to advertise 500 stuffed teddy bears for the holiday season. They will retail for $18.00 each. She already has 114 teddy bears, which cost $10.50 each, in the stock room. What is the most she can pay for each remaining teddy bear if she wants to achieve a departmental markup of 54%?

100%

	Units	Cost Each	C$ Total	Retail Each	R$ Total	MU%	C%
Dept. goals and/or plans	_____	_____	_____	_____	_____	_____	_____
On hand	_____	_____	_____	_____	_____	_____	_____
Open to buy	_____	_____	_____	_____	_____	_____	_____

5. A sporting-goods buyer needs a 43% markup on his merchandise. He buys 12 dozen baseball hats at $36.00 per dozen and retails them at $5.00 each. He then buys 20 dozen team-monogrammed baseball hats that cost $4.75 each. At what price will he mark the 20 dozen pieces in order to achieve his overall markup on the entire order?

100%

	Units	Cost Each	C$ Total	Retail Each	R$ Total	MU%	C%
Dept. goals and/or plans	_____	_____	_____	_____	_____	_____	_____
On order	_____	_____	_____	_____	_____	_____	_____
On order	_____	_____	_____	_____	_____	_____	_____

Solve these problems without setting up the charts. Think through what steps you need to take.

6. A buyer purchases the following items:

12 hand-weight sets at $4.75 to retail for $10.00
24 10-pound weights at $6.00 to retail for $12.00

She also bought 48 steppers for the exercise department at $7.75 each. What is the most competitive price she can mark the steppers and still achieve an average markup of 45% on this order?

7. A buyer needs $160,000 at retail value in stock. The departmental markup is planned at 52%. The buyer places an order for $54K cost and $106K retail. What are the remaining cost and retail orders to be placed, and what are the markup dollars and percent on the balance?

8. While planning a music promotion, a buyer plans for 400 CDs in the department. The CDs are to retail at $15.00 each. The buyer purchased 100 titles for $9.00 each. The departmental markup is 47.5%. What are the balance of cost dollars to be spent in all and the individual cost?

9. The infantwear buyer designed rompers for her department that she would bring in at a wholesale price of $96.00 per dozen and a coordinating T-shirt at $72.00 per dozen. She brings in 30 dozen rompers and 40 dozen T-shirts. She plans to mark the rompers at $16.00 each. At what does she have to mark the T-shirts to achieve an average markup of 52%?

10. The small-leather-goods buyer for Sue and Chrissy's Gift Shoppe bought a closeout batch of 300 small leather pieces for $2,400.00 cost. If she prices 100 pieces at $20.00 each and 75 pieces at $10.00 each, what is the minimum price at which the remaining pieces have to be marked to maintain the average markup of 48%? At what retail price do you think the buyer would set the pieces?

Cumulative Markup

11. A buyer was quoted the following prices on knit tops during a manufacturer's closeout sale. In addition to the low selling price, the buyer negotiated a 10% additional seasonal discount. All tops are to be retailed at the same price. What retail price will result in a 52% markup?

    100 tops at $18.00 each
    150 tops at $15.00 each
    250 tops at $14.00 each
    300 tops at $17.00 each

First determine the total cost price and then the total retail value. What will the individual retail price be to earn a 52% markup?

	Dollars	Percents
R =		100%
MU =		
C =		

Reminder: Once you find the total retail, divide the total dollars by the total number of pieces to determine the individual price.

Individual retail price: _____

12. The following dresses were purchased by a buyer:

25 pieces at $29.00 each
30 pieces at $32.00 each
22 pieces at $35.00 each

The average markup must be 50%. What is the retail price for each dress if all the dresses are to be sold at the same price?

	Dollars	Percents
R =		100%
MU =		
C =		

Reminder: Once you find the total retail, divide the total dollars by the total number of pieces to determine the individual price.

Individual retail price:_____

Solve these without a chart.

13. A buyer is offered a group of windbreakers for boys that have all the national basketball team logos. Some of the windbreakers are hooded, and some are not. The manufacturer will sell 165 windbreakers at a cost of $25.00 each and 200 hooded windbreakers at a cost of $28.00 each. The buyer plans to sell them all at the same retail price. The departmental markup is 43%. What must the retail price be for all the windbreakers to achieve that goal?

14. A buyer for a gift store purchased some pewter and glass gift items. The assortment included the following:

    10 cat figurines at $6.00 each

    12 candlesticks at $14.00 each

    15 pewter-bottom, glass-top boxes at $10.00 each

    11 serving bowls at $9.50 each

    She plans to sell all the gift items at the same price. What unit retail will result in a 54% markup?

15. A buyer purchased a girls' 4–6X lingerie closeout consisting of 275 flannel nighties at $40.00 per dozen and 475 cotton-knit night shirts at $3.25 each. He wants to retail all this merchandise at the same price. If he needs a 52.5% markup on this purchase, what is the unit retail price? What price should the buyer promote them for, and why?

Maintained Markup

16. A buyer purchased 200 pairs of soccer shin guards at $4.50 each and retailed them for $8.00 each. After 100 pairs were sold at the original retail price, the remaining pieces were marked down to $6.00 each. At this price all the shin guards were sold. First determine the initial markup percent and then compare the results to the maintained markup percent.

Initial Markup

	Dollars	Percents
R =		100%
MU =		
C =		

Maintained Markup

	Dollars	Percents
R =		100%
MU		
C =		

Why would a buyer want to analyze this problem? What could it tell the buyer?

17. Sixteen dozen pairs of earrings were bought for the preteen girls' accessory department at a cost of $0.75 each and were priced at $1.25 per pair. One hundred seven pairs were sold at $1.25, and the remaining pairs were marked up to $1.75. At a later date, 32 pairs remaining in stock were reduced to $0.95 and then sold. First determine the initial markup percent, and then compare the results to the maintained markup percent.

Initial Markup

	Dollars	Percents
R =		100%
MU		
C =		

(*Hint:* In order to find maintained markup easily, simply take the total number of pieces bought and calculate the retail earnings at each price point. There were 16 × 12 = 192 pieces. Find how many sold at $1.25, how much was earned then, how many were sold at $0.95 and how much was earned, and, finally, how many were sold at $1.75 and the earnings. Then just fill in the maintained-markup chart.)

Maintained Markup

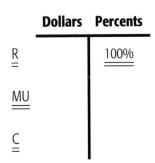

	Dollars	Percents
R =		100%
MU		
C =		

Why would a buyer want to analyze this problem? What could it tell the buyer?

18. A sporting-goods buyer purchased 1,500 NHL hockey jerseys at $24.00 each for a special new-store-opening promotion. They were advertised at a $40.00 retail, and 1,000 jerseys were sold. Then the shirts were mixed in with regular stock at $50.00 each. Between July 1 and September 12, all but 50 pieces were sold at this price. The remaining 50 pieces were reduced to $25.00 and all but 10 were sold. Those 10 were sold during a midnight madness special for $9.99. Determine the initial markup percent, and then compare the results to the maintained markup percent.

Initial Markup

	Dollars	Percents
R =		100%
MU =		
C =		

In order to find the maintained markup easily, take the total number of pieces bought and calculate the retail earnings at each price point.

Maintained Markup

	Dollars	Percents
R =		100%
MU =		
C =		

Why would a buyer want to analyze this problem? What could it tell the buyer?

19. 200 pairs of jeans were bought for a young men's department to be promoted at $12.99. The jeans were purchased for $7.00 per pair with discounts of 10/15. On the day of the ad release, 142 pairs

were sold. The remaining pairs were marked into regular stock for $18.00 and within the month all were sold. Using this information determine the cost of the merchandise, the initial markup, and the maintained markup. (*Hint:* You will have to deduct the series discounts to determine the cost dollars.)

Initial Markup

	Dollars	Percents
$\underline{\underline{R}}$		$\underline{\underline{100\%}}$
$\underline{\underline{MU}}$		
$\underline{\underline{C}}$		

Now, based on the selling results, calculate the maintained markup.

Maintained Markup

	Dollars	Percents
$\underline{\underline{R}}$		$\underline{\underline{100\%}}$
$\underline{\underline{MU}}$		
$\underline{\underline{C}}$		

What do these figures tell the buyer?

20. A buyer for the home-furnishings area tries to purchase merchandise at a 50% markup or more initially, because she must promote heavily in her department. The overall department markup is 48%. Determine how this purchase falls into the buyer's goals. One hundred bed-in-the-bag sets were purchased for $20.75 to be promoted at $49.99. An additional 4 dozen comforters were purchased at $14.75 each to be promoted at $29.99. Eighty bed-in-the-bag sets sold at $49.99, and 28 comforters sold at $29.99. At the end of the month the buyer took an additional $10 markdown on the remaining pieces and transferred them to a special branch-store sale. The remaining pieces all sold at the new retail price. First determine the initial markup percent, and then compare the results to the maintained markup percent.

Initial Markup

	Dollars	**Percents**
R		100%
MU		
C		

Now, based on the selling results, calculate the maintained markup.

Maintained Markup

	Dollars	**Percents**
R		100%
MU		
C		

What could these figures tell the buyer about the profitability of using this vendor to achieve the departmental goal?

six

Markdowns

In Chapters 4 and 5 you learned how to price merchandise. Initially, buyers focus on individual pieces and initial markup. To meet financial goals, a merchandiser will determine the average markup of a purchase and compare that to the departmental markup. Cumulative pricing for higher profits and reviewing the overall maintained markup are essential as buyers evaluate the sales performance of the products they purchased. In this chapter you will see how merchandisers make price adjustments that either increase or decrease the retail selling price to encourage sales. Also, you will see that the price changes must always be recorded in order to maintain the following:

1. An accurate accounting or book inventory in line with the physical value of the stock.
2. A means of controlling and managing the amount of stock to be decreased in value in order to maintain profit goals. A 6-month merchandising plan incorporates markdown reductions to stock in the financial planning and evaluation process, because when the stock is reduced in value, the buyer has more money to use to buy new product.
3. Repricing accounting also allows the retailer to plan and adjust the departmental markup goals. Repricing means that sometimes the product price can be increased, as was demonstrated with maintained markup problems.

The most common type of repricing is the **markdown,** which is a reduction in retail price. Keep in mind that when the prices of products are reduced, the value of the inventory, or stock on hand, is also reduced. Buyers are keenly aware of markdowns as they develop marketing strategies to meet sales goals for a department or store. It is important to note that markdowns are not bad; instead, they serve as a tool that allows buyers to move out stock. When old stock is sold, the buyer can purchase more product and offer the consumer new choices.

Typical reasons for reducing the price of inventory include the following:

1. Buying errors—i.e., wrong sizes or colors or poor timing
2. Pricing errors—i.e., prices too high, competition has lower prices
3. Promotions
4. Broken assortments
5. Poor sales service
6. Poor display work
7. Decreasing consumer demand, but merchandise can still be sold at a lower price
8. Covering shortages and giving employee discounts

Good judgment is invaluable in determining when to begin the new selling price as well as in deciding how much of a price change should be taken.

Traditional Markdown

The following examples show how markdowns are calculated on individual items. The traditional markdown is taken as follows:

$$\text{Old price} - \text{new price} = \text{difference}$$

The difference is known as the *dollar markdown.*

$$\text{Dollar markdown} \times \text{number of pieces} = \text{markdown dollars}$$

E X A M P L E

A buyer reduces 68 pieces of a $20.00 scarf to $14.99. Determine the markdown dollars.

$$\$20.00 - \$14.99 = 5.01$$
$$68 \times \$5.01 = \underline{\$340.68} \text{ markdown dollars} \quad \blacksquare$$

These markdown dollars accumulate each month, and a merchandiser keeps the total dollar values on the 6-month merchandising plan.

Promotional Discount Markdown Reductions

Due to the emphasis on in-store promotions, consumers will often find merchandise advertised with a percent discount:

SKI BOOTS

30% off the ticketed price

Promotional markdowns are calculated as follows:

Current retail price $\times$ percent off = dollar markdown

EXAMPLE

Jeff's Ski Stop advertises 30% off 72 pairs of ski boots that retailed at $100.00 each. All 72 pairs were sold. Find the total markdown dollars. The total markdown for one pair is $100.00 × 30% = $30.00.

$30.00 $\times$ 72 = _$2,160.00_ total markdown dollars ∎

Markdowns are always determined after consideration of the initial markup and the maintained markup of a product. It is important for a buyer to be able to estimate about how many pieces will be sold in such a sale, based on the current consumer buying trends, and then estimate what the markdown dollars will be. A buyer must make solid judgments when putting merchandise out for special promotions because, although the buyer wants to encourage business, the products still need to have enough markup to contribute to expenses and profitability. Markdown dollars accumulate each month, and merchandisers find the total dollar values when doing the 6-month merchandising plan.

Employee Discounts

An employee of a retail organization usually earns a discount on merchandise. This is another type of price reduction, and it must also be recorded because it lowers the value of the store inventory. This reduction is included in the dollar markdowns shown on the 6-month merchandising plan.

EXAMPLE

Employees receive a 20% discount on all merchandise. If an employee buys a top at $28.00, the employee discount is $28.00 × 20% = _$5.60_.

$28.00 − $5.60 = _$22.40_

The employee allowance is $5.60, and the employee pays $22.40 plus tax. ∎

Employees sometimes can earn multiple discounts, but these discounts are not combined. Consider the following example.

EXAMPLE

The housewares department of Peter's Pottery Store is offering 30% off all food processors during a special 1-day sale. A store employee chooses a model that retails at $159.99. The employee receives the 30% in-store promotion plus an additional 20% employee discount. How much will the employee pay for the food processor?

You can solve with your calculator.

159.99 ⊟ 30 % ⊟ 20 % 89.5944

159.99 ⊟ 89.59 ⊟ 70.4 markdown dollars

The markdown dollars are $70.40.

Note that you subtract 30% of $159.99 and then 20% of that result, or 20% of $111.99. You do not combine the percentages and take 50% of $159.99 to find the discount. ∎

■ Problems

Traditional and Promotional Markdowns

1. For Valentine's Day a buyer offered 30% off for red and white sweaters priced at $100.00 each. Sixty sweaters were sold.

   a. Determine the new selling price.

   b. Determine the total amount of markdown dollars taken.

2. A buyer reduced ski jackets for boys and girls in November. There were 112 priced at $55.00 each, and the buyer reduced them to $36.99.

   a. What was the percentage reduction? (Think back to the *Helpful Hints*.)

   b. How much in markdown dollars was taken on each piece?

   c. What was the total markdown taken?

3. Ben, an employee of Chris's Surf Shop, purchased a new surf board. The board retailed at $125.00. Ben earns a 20% discount as an employee. What was the price of the surf board?

4. A special discount was offered to the employees of Ryan's, a leading men's store. Employees were given their employee discount of 35% off and an additional 15% discount on all purchases. The following merchandise was sold:

   15 cable knit sweaters, which retailed at $65.00 each

   22 pairs of khaki twill pants, which retailed at $50.00 each

   18 casual jackets, which retailed at $62.00 each

   a. What was the total sales volume?

   b. Find the total markdown dollars taken.

 Hint: Remember what you learned in Chapter 3 about series discounts.

5. An area merchandiser reduced 62 pairs of ladies' shorts from $40.00 to $32.99 for a sale. After the sale 26 pairs remained. Those pieces were then marked down to $24.99.

   a. Find the total markdown amount taken.

   b. What was the percent deduction from the original amount of $40.00 to the final markdown price of $24.99? (Think back to the *Helpful Hints*.)

6. A tote bag retailing for $79.00 was advertised at 30% off for a special sale. What was the new selling price?

7. Thirty-six sofas were sold during a clearance sale at Boy's Furniture. All 36 sofas were originally marked $999.00 and sold at a clearance price of $679.00.

   a. Find the total markdown amount taken.

   b. What was the percentage deduction from the original amount of $999.00 to the markdown price of $679.00?

8. Mo's Petite Shop ran an ad for 25% off all accessories. The following items were sold:

   15 wallets originally priced $25.00 each

   8 woven belts originally priced $20.00 each

   60 pairs of earrings originally priced $12.00 each

   a. What was the total sales volume?

   b. Find the total markdown dollars taken.

9. A men's buyer had 120 sport coats priced at $175.00. Sales were not good, and after only 20 sport coats had been sold, the buyer reduced them to $124.99. At the new price 36 coats were sold. The buyer then took an additional markdown to $87.99.

   a. Determine the original markdown dollars taken.

   b. Find the total markdown dollars taken.

10. The rug department was running a special sale during an annual home furnishings' promotion. All area floor rugs were advertised at 40% off. Sales for the promotion were

    40 dhurries originally marked $295.00   New price _____

    25 berbers originally marked $175.00   New price _____

    12 patterned needlework rugs originally marked $115.00 New price _____

    a. Find the new selling prices for these rugs.

    b. Find the total markdown dollars.

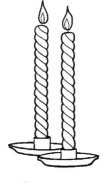

11. A gift buyer for the Gift Spot bought 200 pairs of brass candlesticks and marked them $25.00. The buyer then placed them on sale during a special 3-day sale for 20% off. During the sale 162 pairs were sold. The remainder were marked back to the original price and were sold out at $25.00.

    a. Determine the total markdown dollars taken.

    b. Determine the total retail dollars earned on all 200 pairs of candlesticks.

12. A buyer purchased 300 pieces of silver jewelry off season that all originally retailed at $15.00 each. During a Mother's Day sale the jewelry was 25% off. All but 23 pieces were sold. Determine the total markdown dollars.

13. The Comedy Shop was selling old comic books at 20% off. Chris purchased comic books with a retail value of $68.00.

    a. What was the actual price that Chris paid for the comic books?

    b. Find the markdown dollars on this purchase.

14. Bob Sause's Cooking Gallery took the following markdowns:

    18 omelet pans originally priced at $15.00, new price $9.99

    22 ceramic colanders originally priced at $20.00, new price $12.99

    17 cookbook holders originally priced at $10.00, new price $5.99

    16 copies of Lea's Delights, a dessert cookbook originally priced at $15.00, new price of $9.99

    Find the total markdown taken on this merchandise.

15. A sporting goods' buyer advertised the entire stock of roller blades for a 6 P.M. to 9 P.M. 1-day-only sale at 25% off. At the end of the promotion the following sales were determined:

    38 pairs of Lightning blades originally $219.00 per pair

    46 pairs of BB Turbo Trick blades originally $269.00 per pair

    17 pairs of MM hockey blades originally $119.00 per pair

    62 pairs of TR blades originally $179.00 per pair

    a. Find the new selling prices for these styles.

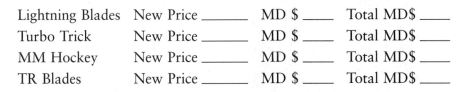

Lightning Blades	New Price _____	MD $ ____	Total MD$ ____
Turbo Trick	New Price _____	MD $ ____	Total MD$ ____
MM Hockey	New Price _____	MD $ ____	Total MD$ ____
TR Blades	New Price _____	MD $ ____	Total MD$ ____

    b. Find the total markdown dollars taken._____

16. The Styles Shop ran a week-long 15%-off sale on all custom monogrammed stationery. At the end of the week, sales came to $8,540. What was the original retail value of this merchandise? (Think back to the *Helpful Hints*.) Find the markdown dollars taken.

17. A special 3-day 25%-off sale for scarves was held during the holiday season. Use the chart to determine prices and total markdown dollars. Calculate the markdown price to the nearest $0.99.

Units	Original Price	New Price	Markdown Difference	Units Sold	Total MD$
36	$65.00	_____	_____	23	_____
60	$45.00	_____	_____	37	_____
72	$40.00	_____	_____	52	_____
$2\frac{1}{2}$ doz.	$75.00 each	_____	_____	17	_____
Total					_____

Determine the new price and the total markdown dollars.
(*Remember*: Round off to $0.99. For example, use your calculator to determine the new selling price:

$$75 \boxed{-} \: 25 \boxed{\%} \qquad \boxed{56.25}$$

Rounded to the nearest $0.99, $56.25 is $55.99. This way you are selling at 25% off or less. If you marked the item $56.99, you would be engaged in false advertising, because $56.99 is not less than 25% off.

Merchandise Planning
The following problems are helpful in analyzing plans and goals established in merchandise planning and goal setting. In this section you will see how markdown dollars are compared to the sales volume generated. These comparisons are made by using the same basic techniques established in Chapter 2.

Helpful Hint #1:
   To find markdown dollars: net sales × MD% = markdown dollars

Helpful Hint #2:
   To find markdown percentage: markdown dollars ÷ net sales = MD%

Helpful Hint #3:
   To find net sales: markdown dollars ÷ markdown % = net sales

18. A handbag buyer took markdowns in the month of February totaling $1,600.00. During the month the net sales for the department were $17,500.00. The original markdown plans called for 12% in markdowns.

    a. Determine the number of markdown dollars originally planned.

    b. Were the actual markdowns taken over or under the plan?

    c. By how many dollars were they over or under?

19. A buyer bought 200 wallets and retailed them at $25.00 each. At the end of the season 40 wallets were sold in a clearance sale at one-half off.

    a. Find the total markdown dollars.

    b. What was the retail value of all the merchandise sold?

20. The exercise and activewear buyer has estimated sales for the spring season to reach $650,000.00 with markdowns at 10.5%. At the end of the season the buyer had taken $57,250.00 in markdowns.

    a. Was the actual markdown over or under the plan?

    b. Specifically, how many dollars over or under the plan was the actual markdown?

21. A buyer took monthly markdowns in the children's area as follows:

    50 dresses from $20.00 to $15.00

    30 sweatshirt sets from $25.00 to $20.00

    24 sweaters from $12.00 to $8.00

    The gross sales for the month were $115,000.00. Customer returns and allowances totaled 7.3%.

    a. Determine the net sales on the merchandise sold.

    b. Determine the actual markdown in dollars.

    c. Determine the monthly markdown percentage in relation to the monthly net sales.

Don't forget your *Helpful Hints*.

22. A buyer had 110 raincoats in stock at a retail price of $175.00. The raincoats were reduced to $114.99. Sales for the month were planned at $195K. Markdowns were planned at 5% of the planned sales.

   a. Determine the markdown dollars taken on the raincoats.

   b. Determine the planned departmental markdowns.

   c. Did the buyer have enough markdown money planned to allow reduction of the raincoats?

23. During a friends and family in-store promotion, the local specialty store sold 36 denim work shirts, which are priced at $45.00, and 23 flannel shirts, which retail at $36.00, to employees and their family members. A discount of 35% was offered.

   a. Determine the sales volume achieved on these shirts.

   b. Find the markdown dollars taken.

24. At the end of inventory clearance Aaron went shopping in the men's department. He bought

   3 pairs of pants at $48.00 that were selling at 30% off

   2 shirts at $25.00 each that were reduced by 35%

   1 jacket retailing at $165.00 that was selling for 40% off

   In addition to the sales discounts, Aaron earned an employee discount of 20%.

   a. What was the original retail value of the merchandise?

   b. How much did Aaron pay for all the merchandise?

Hint: Remember what you learned in Chapter 3 about series discounts.

25. For a special sale, the shoe-department buyer purchased two styles of sandals at a negotiated cost of $16.00 each. The buyer decided to retail 144 pairs at $36.00 and the other 96 pairs at $32.00 each. The following are the selling results:

144 pairs originally retailed at $36.00
24 pairs sold at $36.00
25% off markdown was taken, 84 pairs sold
Sandals marked to $17.99, all sold

96 pairs originally retailed at $32.00
32 pairs sold at $32.00
25% off markdown taken, 46 pairs sold
Sandals marked to $14.99, all sold

a. What was the initial markup in this purchase?

b. What was the maintained markup on this purchase?

c. Find the total markdown dollars taken on this purchase.

Initial Markup		
	Dollars	**Percents**
R		100%
MU		
C		

Maintained Markup		
	Dollars	**Percents**
R		100%
MU		
C		

Merchandise Planning

seven

Elements of a Six-Month Merchandising Plan

Purpose of a Six-Month Plan

All stores, small and large, benefit from organized planning. When working with set goals, profits are more likely. The major tool in merchandising for guiding merchants toward desired sales goals and stock assortments is the six-month dollar merchandising plan. The six-month plan is a road map covering 26 weeks of the year that enables a buyer or manager to effectively

1. Plan and evaluate sales.
2. Plan and evaluate stock, including stock-to-sales ratios, average stock, and turnover.
3. Plan and evaluate stock reductions.
4. Plan and evaluate the flow of purchases.

With the ability to plan carefully and effectively, it also allows the buyer or manager to

1. Establish goals.
2. Provide a plan to follow in order to reach the goals.
3. Provide a means of measuring results and analyzing for future improvement.

Ultimately the success of an operation will depend on how carefully a buyer follows the plan that controls the amount of money spent to produce effective sales without overstocking or understocking the store.

The following terms and phrases are used daily in the merchandising industry. Each concept defined is involved in the use of a six-month plan. Take some time to review these words, and then you will go through each component of a six-month plan and see how the terms apply.

Industry **Terms** and *jargon*

Six-Month Plan   A form used to coordinate a budget that sets sales goals and determines the amount of merchandise necessary to reach desired goals. See Figure 7.1.

Spring Season   The period of February 1 through July 31.

Fall Season   The period of August 1 through January 31.

LY   Last year.

TY   This year.

Actual   Specific results that actually occur.

Revised   A new version of an original plan.

Workroom Costs   A cost incurred to complete the sale that is not planned for in the initial markup. Examples include hemming bridal gowns and altering men's suits (also identified on operating statements (see Chapter 13)).

Cash Discounts   Monies earned by paying invoices within the negotiated dates, expressed as percentages and dollars (also identified on operating statements (see Chapter 13)). (Review Chapter 3.)

Stock Turnover   Also referred to as **turnover,** a ratio that represents the number of times merchandise is sold and replaced in a specific period of time. Turnover rate is often called a *checking tool.*

Shortage/Shrinkage   When the physical inventory is less than the financial inventory records. The financial records are called the *book* (accounting records) *inventory* (also identified on operating statements (see Chapter 13)).

Overage   When the physical inventory is greater than the book inventory (also identified on operating statements (see Chapter 13)).

Gross Sales   All sales made in a specific period of time (also identified on operating statements (see Chapter 13)). (Review Chapter 2.)

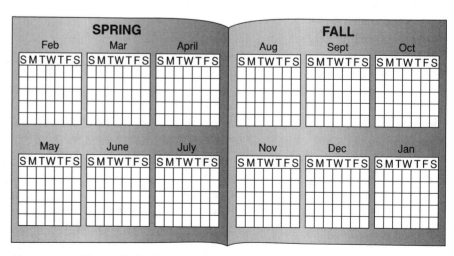

Figure 7–1   Six-month planning calendars.

Customer Returns and Allowances   Credits issued for merchandise brought back after a sale (return) or a credit allowed for a defect or price adjustment (allowance) (also identified on operating statements (see Chapter 13)). (Review Chapter 2.)

Net Sales   Total final sales for a period of time. Gross sales minus customer returns and allowances equals net sales (GS 2 CRA = NS).

Sales Volume   Same as net sales.

Season Sales   Same as net sales.

Sales Forecasting   Projecting future sales, based on consumer and economic trends.

Average Stock or Average Inventory   The general amount of inventory on hand at any one given time.

BOM Stock   Beginning-of-the-month stock.

EOM Stock   End-of-the-month stock. (Note: The end of one month is the same as the beginning of the next, i.e., February EOM = March BOM.)

EOS Stock   End-of-the-season stock. This stock figure must be low enough to clean out old-season stock and yet strong enough to present a good merchandise assortment for the next season. EOS stock is January 31 for the fall season and July 31 for the spring season.

Stock-to-Sales Ratio (SSR)   The ratio of stock on hand to monthly sales. This is often called a planning tool. Stocks need to peak prior to major selling seasons in order to offer customers a strong assortment.

Stock, Inventory, Merchandise, Goods   Interchangeable terms.

Physical Inventory   The specific amount of merchandise, counted at least two times a year (January and July).

Book Inventory   The financial bookkeeping records of the inventory based on the dollar value of the physical count plus purchases minus sales, stock reductions, and employee discounts. This term is applied in operating statements (also identified on operating statements (see Chapter 13)).

Monthly Markdown Percent   Percent value comparing monthly dollars of reduced stock to monthly sales. (Review Chapter 6.)

Markdown Percent for Season   Percent value comparing total season dollars of reduced stock to total season sales.

Annual   A 1-year period of time from January 1 through December 31.

Fiscal Year   An accounting period of 12 months. In the merchandising market, the fiscal year runs from February 1 through January 31.

Retailer's Calendar   Also called a 4-5-4 calendar, a calendar used to give a selling season the same number of selling days so businessmen and women can compare performance. Traditionally, the

4   5   4

S	M	T	W	T	F	S
1	2	3	4	5	6	7
8	9	10	11	12	13	14
15	16	17	18	19	20	21
22	23	24	25	26	27	28

S	M	T	W	T	F	S
29	30	31	1	2	3	4
5	6	7	8	9	10	11
12	13	14	15	16	17	18
19	20	21	22	23	24	25
26	27	28	29	30	1	2

S	M	T	W	T	F	S
3	4	5	6	7	8	9
10	11	12	13	14	15	16
17	18	19	20	21	22	23
24	25	26	27	28	29	30

Figure 7-2   4-5-4 calendar.

calendar begins on February 1 and shows 4 weeks, then 5 weeks, and, finally, 4 weeks. Thus, each segment of the year has 13 weeks, which equal one quarter. See Figure 7.2.

Quarter   A 13-week period. There are four quarters in a year. At the end of a quarter, a business examines the financial results for that period and compares those figures to last year's and to consumer and economic trends.

Merchandise manager   The person responsible for the merchandise that a store sells. The MM has a strong knowledge of the store customer and is responsible for seeing that the merchandise assortment addresses the company's target market.

Controller   The person responsible for the financial planning for a company, for establishing financial goals, and for maintaining profitability.

Elements of a Six-Month Plan

A six-month merchandising plan is the form used to coordinate sales and merchandising goals in a budget. Although there are many styles of forms for a six-month plan, the information is the same on all the different forms. The following identifies the columns and rows on a six-month plan and explains the purpose of each.

A Department name.

B Identification number (important for computerized planning).

C This year's goals.

D Last year's results.

E Costs incurred to complete a sale (for both LY and plan). Important for profit-and-loss (P/L) statements.

F Cash discounts earned (for both LY and plan). Important for P/L statements.

G Shortage percentage (LY/plan). Important for P/L statements.

H Average stock or average inventory at retail value (LY/plan).

I Season stock turnover (take out two places to the right of the decimal point.) (LY/plan).

J Total 6-month markdown percentage (LY/plan).

K Departmental average markup. (LY/plan). (See Chapter 5.)

L Two seasons shown, identifying specific months.

M Net sales for the 6-month period.
- A section identifies the percentage of sales per month. This row is used during analysis or planning calculations.
- Total season sales are shown under season total.
- Each month for last year, the plan, the revised plan, and actual results are identified.

N Beginning-of-the-month stock totals, identifying last year, the plan, the revised plan, and the actual results. Remember, the beginning of one month is the same as the end of the previous month. There is not a total stock figure. (See R.)

O Monthly stock reductions (see Chapter 6) for the last year and plan.
- Monthly percentage to sales for the last year and plan are given.
- The total markdowns for the 6 months are shown under season total, and the percent is shown with the overall control data also.

P Monthly retail purchases, identifying last year, the plan, and the revised and actual results. The total purchases at retail are identified under the season total for the 6 months.

Q Monthly purchases at cost are identifying last year, the plan, and the revised and actual results. The total purchases at cost are identified under the season total for the 6 months.
(Review Chapter 4: $100\% - MU\% = C\%$ and $R\$ \times C\% = C\$$)
Use MU% identified in K.

R July 31 (same as August 1) or January 31 (same as February 1) end-of-season stock figure.

S Monthly stock-to-sales ratio. This section, just like the sales percent each month, is used when analyzing and developing the plans (LY/plan).

T Signatures of the buyer, who prepares the plan, the merchandise manager, who approves the plan, and the controller, who approves the plan and combines these figures with those of all the other departments to determine the operations of the total company.

	Department Name:	A	No. B	
		Plan (This Year) C	**Actual** (Last Year) D	

SIX-MONTH MERCHANDISING PLAN

	Plan (This Year)	Actual (Last Year)
Workroom Cost	E	
Cash Discount %	F	
Shortage %	G	
Average Stock	H	
Season Stock Turnover	I	
Overall Markdown %	J	

Departmental Markup %

Last Year	Plan	Actual
K		

		Feb.	Mar.	Apr.	May	June	July	**Season Total**
Spring 19- L		Feb.	Mar.	Apr.	May	June	July	**Season Total**
Fall 19-		Aug.	Sept.	Oct.	Nov.	Dec.	Jan.	M
Sales Percent Each Month		M						
Sales $	Last Year	M						
	Plan							
	Revised							
	Actual							
Retail Stock (BOM) $	Last Year							
	Plan	N						
	Revised							
	Actual							
Markdowns	Last Year							
	Plan (dollars)							
	Last Year Monthly %	O						
	Plan Monthly %							
	Actual Dollars							
Retail Purchases	Last Year							
	Plan	P						
	Revised							
	Actual							
Cost $	Last Year							
	Plan	Q						
	Revised							
	Actual							
Ending Stock July 31 Jan 31	Last Year							
	Plan	R						
	Revised							
	Actual							
Stock to Sales Ratio	Last Year	S						
	Plan							

Department Buyer _____ T _____   Merchandise Manager _____

Controller _____

■ Problems

1. What two periods does a six-month merchandising plan cover?

2. How many quarters are in each season?

3. When looking for the previous year's performance on the six-month plan, you will look for the actual figures. More commonly, they are called _____ and abbreviated as _____ .

   When looking for the future year's performance on the six-month plan, you will look for the planned figures. More commonly, they are called _____ and abbreviated as _____ .

4. Why is the revised row an important factor on a six-month plan?

5. Identify three external factors that might cause a plan to be revised.

6. Identify three internal company factors that might cause a plan to be revised.

7. How does a holiday season affect a six-month plan's sales forecast?

8. Why does the merchandising industry work on a fiscal year beginning February 1 rather than the annual year beginning January 1?

9. Why are sales the first area to be identified on a six-month merchandising plan?

10. If the stock-to-sales ratio is 3 to 1, what does that mean?

11. If a department has a 2.0 stock turnover for a 6-month period, what does that mean?

12. Explain what it means if the merchandise manager says the turnover is too slow.

13. What happens to the merchandise if the turnover rate is too slow?

14. Can a turnover rate be too fast? Why?

15. What does a fast turnover rate do for consumer loyalty?

16. Why is it important to identify markdown reductions on a six-month plan?

17. Why is it important to identify the purchases each month on a six-month plan?

18. The EOS for spring is _____ .

    The EOS for fall is _____ .

    The BOM for April is the same as the EOM for _____ .

    The BOM for July is the same as the EOM for _____ .

    The BOM for August is the same as the EOM for _____ .

    The BOM for January is the same as the EOM for _____ .

    The BOM for October is the same as the EOM for _____ .

19. If you know the retail dollars purchased and you know the departmental average MU%, using the formulas already covered, how would you determine the cost percent of purchases?

20. Why is it important for three different people to sign their approval of a six-month plan?

eight

CHAPTER

Analyzing Last Year's Merchandising Plan

Every 6 months buyers are required to analyze the financial operations of their departments. This chapter focuses on analysis of the financial plans, allowing the merchant to interpret the fluctuations in stocks and sales of the department. In this chapter we go through the same step-by-step process a buyer takes to review the statistics for a 6-month period.

Step 1: Analyze Sales

Buyers always look first at the sales. Not only do they want to know what the sales are, but they also want to know how the dollars were earned over the 6-month period. This type of charting visually demonstrates peak months and sales trends and quickly points out selling slumps. On the 6-month plan, you work with sales volume, or net sales.

Remember: Gross sales – customer returns and allowances = net sales.

To evaluate sales: Monthly sales ÷ total sales = monthly sales %. The percents are added. The total of the 6 months' percents must be 100%.

Remember: This is just like Helpful Hint 2.

Step 2: Evaluate Stock

Evaluating stock is important to maintain balance. True, stock varies due to the frequency of purchases; however, stores can dictate the stock turnover by buying small amounts often for a quick turnover or by buying large amounts less frequently for a slow turnover. There is no policy that is always right; therefore, reviewing the stock-turnover rate provides a way to *check* the operation, and the stock-to-sales ratio provides a way to balance, or *plan,* the operation. These formulas are essential in future planning when you discuss lead time, reordering, and basic stock requirements. First it is important to identify the SSR to determine the ratio of stock on hand to monthly sales.

To evaluate the stock-to-sales ratio:

First: BOM stock ÷ monthly sales = monthly stock-to-sales ratio

Step 3: Evaluate Markdowns

Reductions in stock reduce the retail value of the inventory. These reductions must be anticipated, estimated, and included when planning. By establishing a plan based on current market trends, along with past information from both the department and the NRF'S FOR Annual Report, a buyer is able to control stock reductions realistically. Each month may still fluctuate, but by comparing both the monthly markdowns to monthly sales and the total markdowns to the overall season sales volume, a buyer can stay on top of the reductions. Remember, markdowns are the most common type of price change and can act as an effective tool for increasing sales and providing room for new merchandise.

To evaluate markdowns:

First: Monthly markdowns ÷ monthly sales = monthly markdown %

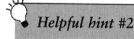

Helpful hint #2

Second: Total markdown dollars for 6 months ÷ total sales
= total markdown %

Note: The total markdown percent figure will go on the top of the six-month merchandising plan with the average stock and turnover data.

Step 4: Evaluate Purchases at Retail

A continual flow of merchandise in a store is essential for a profitable operation. By timing merchandise purchases, a buyer is able to maintain a good flow of stocks and sales, therefore keeping the assortment exciting and always changing.

To evaluate monthly purchases at retail:

Monthly sales + EOM stock + monthly markdowns − BOM = purchases

Another way of calculating how much stock is needed to meet the monthly stock levels is

BOM − monthly sales − monthly markdowns − EOM = purchases to restock

(Don't forget to add the 6-months figures and fill in the totals.)

Step 5: Evaluate Purchases at Cost

Buyers and controllers also like to look at the cost of the merchandise. This figure quickly points to the dollars needed to purchase product. It is also a figure that is used on operating statements to identify the cost of the goods purchased.

To evaluate monthly purchases at cost:

$$100\% - \text{departmental MU\%} = \text{cost\%}$$
$$\text{Purchases} \times \text{cost\%} = \text{cost\$}$$

This is like what you did in Chapter 4. $R\$ \times C\% = C\$$

Step 6: Measure Average Stock and Turnover

The final step in evaluating a department's performance is to measure how the sales and stocks balance. The turnover rate will tell if the merchandise is moving too slowly or too quickly. You will also be able to judge when stocks are heavy or light and if markdowns might be needed to reduce any stocks that are turning too slowly. In order to calculate the turnover rate, first you need to know what the average stock on hand is at any one time. Once that figure has been calculated, the turnover rate can be determined.

To evaluate monthly average stock:

$$(6 \text{ BOM} + 1 \text{ EOM}) \div 7 = \text{average stock}$$

To evaluate stock turnover rate:

$$\text{Net sales} \div \text{average stock} = \text{turnover}$$

Using a Six-Month Plan

Let's look at a completed six-month plan and analyze what a buyer or department manager would review. The information is for a men's knit-shirt department, no. 680, for the spring season. Last year this department generated $320,000 in sales. The six-month plan is shown on page 141.

Step 1   The first thing a buyer wants to know is not what the monthly sales were, but what the percent of total sales the month's sales were. There are two reasons: One, the buyer can see the sales pattern, and two, percents provide the most accurate means of comparison. Therefore; for the month of February: $44,800 ÷ $320,000 = 14%. The percentage for each month is calculated in the same way. The sum of the percentages must be 100%.

Step 2   The buyer then wants to review how much stock was on hand to support the sales. To calculate the stock-to-sales ratio for the month of February,

$$\$134,400 \div \$44,800 = 3.0$$

This figure means that there was a 3:1 ratio of stock on hand to sales.

Step 3   The buyer wants to determine the impact of the markdowns each month on the sales, as well as overall. To calculate the monthly markdown percentage for February,

$$\$2,000 \div \$44,800 = 4.46\%$$

To calculate the markdown percentage for the entire season,

$$\$32,000 \div \$320,000 = 10\%$$

Step 4   The buyer then determines the retail purchases for each month. To calculate the purchases for February:

$$
\begin{array}{rl}
\$\ 44,800 & \text{(Feb. sales)} \\
+\ \$133,120 & \text{(EOM)} \\
+\ \$\ \ \ 2,000 & \text{(Feb. markdowns)} \\
-\ \$134,400 & \text{(Feb. BOM)} \\
\hline
=\ \$\ 45,520 &
\end{array}
$$

Note: Complete the retail purchases for each month, and don't forget that the EOM for July is the same as the BOM for August.

Department Name: Men's Knit Shirts No. 680

				Plan (This Year)	Actual (Last Year)
		Workroom Cost			none
SIX-MONTH MERCHANDISING PLAN		**Cash Discount %**			8%
		Shortage %			n/a
		Average Stock			$142,617
Departmental Markup %		**Season Stock Turnover**			2.24

Last Year 44%	Plan	Actual	**Overall Markdown %**		10%

Spring 19- / Fall 19-			Feb. / Aug.	Mar. / Sept.	Apr. / Oct.	May / Nov.	June / Dec.	July / Jan.	Season Total $320,000
Sales Percent Each Month			14%	13%	17%	20%	24%	12%	100%
Sales $		Last Year	44,800	41,600	54,400	64,000	76,800	38,400	$320,000
		Plan							
		Revised							
		Actual							
Retail Stock (BOM) $		Last Year	134,400	133,120	184,960	179,200	176,640	96,000	
		Plan							
		Revised							
		Actual							
Markdowns		Last Year	2,000	2,800	4,200	2,000	12,600	8,400	32,000
		Plan (dollars)							
		Last Year Monthly %	4.46%	6.73%	7.72%	3.13%	16.41%	21.88%	
		Plan Monthly %							
		Actual Dollars							
Retail Purchases		Last Year	45,520	96,240	52,840	63,440	8,760	44,800	311,600
		Plan							
		Revised							
		Actual	25,491.20						
Cost $		Last Year	25,491.30	53,894.40	29,590.40	35,526.40	4,905.60	25,088	174,496
		Plan							
		Revised							
		Actual							
Ending Stock July 31 Jan 31		Last Year	94,000						
		Plan							
		Revised							
		Actual							
Stock to Sales Ratio		Last Year	3.0	3.2	3.4	2.8	2.3	2.5	
		Plan							

Department Buyer _____   Merchandise Manager _____

Controller _____

Step 5 Buyers will also determine the cost of the purchases and review the information seasonally. This information is also found on a profit and loss statement when the inventory for the season is calculated. Last year the departmental markup was 44%. The cost percentage is 56% (100% – 44% = 56%). To calculate the monthly purchases for February,

$$\$45,520 \times 56\% = \$25,491.20$$

Calculate purchases for each month, and don't forget to add the totals for the season.

Step 6 Now it is important to determine both the average stock and the turnover rate. To calculate the average stock, add all 6 BOMs and the end-of-the-season stock and divide by 7.

$$(\$134,400 + \$133,120 + \$184,960 + \$179,200 + \$176,640$$
$$+ \$ 96,000 + \$ 94,000) \div 7 = \$142,617$$

To calculate the turnover rate:

$$\text{Net sales} \div \text{average stock} = \text{turnover}$$
$$\$320,000 \div \$142,617 = 2.24$$

Analysis and Interpretation

The buyer now has the tools to evaluate the past sales performance of this department. By analyzing plans the buyer gains a more realistic outlook on the department's operations. Additionally the buyer can discuss the business trends, fashion outlook, corporate promotions or even the number of selling days in a month, all of which provide a realistic picture of how this department has operated.

■ **Problems**

1. Using the following six-month merchandising plan, fill in last year's figures and then analyze the given information:

   Monthly sales percentage each month

   Monthly stock-to-sales ratio

   Monthly markdown percentage

   Monthly planned purchases at retail and cost

   Average stock

   Stock turnover

   Total markdown percentage

   Last year's spring season figures were as follows:

Sales Volume		Stocks	Markdowns
February	$29,160	$64,152	$1,225
March	$19,440	$50,544	$1,050
April	$22,680	$59,968	$1,850
May	$35,640	$82,050	$2,400
June	$38,880	$74,000	$3,670
July	$16,200	$43,245	$2,795
August	–	$48,000	–

   Departmental markup was 48%.

2. **Analysis and Interpretation** Assume that the store in Problem 1 is a woman's swimwear store located in the southeastern United States in a high tourist area with a large number of other specialty stores. There are new restaurants in this area, and customers are often entertained with music and artists. In your opinion, is this business running a smooth operation? Explain. Discuss the sales flow, the balance of stock, the markdowns taken, and the monthly flow of merchandise coming in as well as the overall figures of markdowns, average stock, and turnover rate.

	Department Name:		No.	
		Plan (This Year)	**Actual** (Last Year)	

SIX-MONTH MERCHANDISING PLAN

	Plan (This Year)	Actual (Last Year)
Workroom Cost		
Cash Discount %		
Shortage %		
Average Stock		
Season Stock Turnover		
Overall Markdown %		

Departmental Markup %

Last Year	Plan	Actual

		Feb.	Mar.	Apr.	May	June	July	**Season Total**
Spring 19-		Feb.	Mar.	Apr.	May	June	July	**Season Total**
Fall 19-		Aug.	Sept.	Oct.	Nov.	Dec.	Jan.	
Sales Percent Each Month								
Sales $	Last Year							
	Plan							
	Revised							
	Actual							
Retail Stock (BOM) $	Last Year							
	Plan							
	Revised							
	Actual							
Markdowns	Last Year							
	Plan (dollars)							
	Last Year Monthly %							
	Plan Monthly %							
	Actual Dollars							
Retail Purchases	Last Year							
	Plan							
	Revised							
	Actual							
Cost $	Last Year							
	Plan							
	Revised							
	Actual							
Ending Stock July 31 Jan 31	Last Year							
	Plan							
	Revised							
	Actual							
Stock to Sales Ratio	Last Year							
	Plan							

Department Buyer _____   Merchandise Manager _____

Controller _____

3. Using the following six-month merchandising plan, fill in last year's figures and then analyze the given information:

Monthly sales percentage each month

Monthly stock-to-sales ratio

Monthly markdown percentage

Monthly planned purchases at retail and cost

Average stock

Stock turnover

Total markdown percentage

Last year's fall season figures were as follows:

	Sales Volume	BOM Stocks	Markdowns
August	$135,000	$405,000	$ 1,050
September	$117,000	$304,200	$ 7,020
October	$108,000	$324,000	$13,450
November	$180,000	$558,000	$14,450
December	$270,000	$513,000	$38,765
January	$ 90,000	$360,000	$29,372
February	–	$325,000	–

Average departmental markup was 51%.

4. **Analysis and Interpretation** The department in Problem 3 is a men's sweater department in a large moderate- to better-priced department store. The location is in the midwestern United States. In your opinion and based on the current styles and trends for menswear, do you feel that this department is running a successful operation? Explain. Discuss the sales flow, the balance of stock, and the markdowns taken monthly and overall; also analyze the average stock along with the turnover rate.

		Plan (This Year)	Actual (Last Year)
	Department Name:		**No.**
	Workroom Cost		
	Cash Discount %		
SIX-MONTH MERCHANDISING PLAN	**Shortage %**		
	Average Stock		

Departmental Markup %			**Season Stock Turnover**		
Last Year	Plan	Actual	**Overall Markdown %**		

Spring 19-		Feb.	Mar.	Apr.	May	June	July	**Season Total**
Fall 19-		Aug.	Sept.	Oct.	Nov.	Dec.	Jan.	
Sales Percent Each Month								
Sales $	Last Year							
	Plan							
	Revised							
	Actual							
Retail Stock (BOM) $	Last Year							
	Plan							
	Revised							
	Actual							
Markdowns	Last Year							
	Plan (dollars)							
	Last Year Monthly %							
	Plan Monthly %							
	Actual Dollars							
Retail Purchases	Last Year							
	Plan							
	Revised							
	Actual							
Cost $	Last Year							
	Plan							
	Revised							
	Actual							
Ending Stock July 31 Jan 31	Last Year							
	Plan							
	Revised							
	Actual							
Stock to Sales Ratio	Last Year							
	Plan							

Department Buyer _____   Merchandise Manager _____

Controller _____

5. Using the following six-month merchandising plan, fill in last year's figures and then analyze the given information:

Monthly sales percentage each month

Monthly stock-to-sales ratio

Monthly markdown percentage

Monthly planned purchases at retail and cost

Average stock

Stock turnover

Total markdown percentage

Last year's spring season figures were as follows:

	Sales Volume	Stocks	Markdowns
February	$20,000	$62,000	$2,002
March	$24,000	$67,200	$1,218
April	$46,000	$92,000	$2,610
May	$36,000	$64,800	$3,218
June	$24,000	$43,200	$3,846
July	$24,000	$85,200	$4,506
August	–	$74,000	–

Average departmental markup was 48%.

6. **Analysis and Interpretation** The department in Problem 5 is a young men's 8–20 better-clothing department featuring blazers and trousers. During the fall this department also supplies local high schools with boys' school uniforms. There are some dress shirts and ties sold as coordinating pieces. In your opinion, after reviewing the sales flow and the balance of stock, do you feel that this department is running a successful operation? Explain. After reviewing the markdowns, do you have any suggestions about the timing markdowns based on times when there would be the strongest flow of natural store traffic, such as the spring holiday and back-to-school selling?

SIX-MONTH MERCHANDISING PLAN	Department Name:		No.	
		Plan (This Year)	**Actual** (Last Year)	
	Workroom Cost			
	Cash Discount %			
	Shortage %			
	Average Stock			

Departmental Markup %			Season Stock Turnover					
Last Year	Plan	Actual	Overall Markdown %					

		Feb.	Mar.	Apr.	May	June	July	Season Total
Spring 19-		Feb.	Mar.	Apr.	May	June	July	Season Total
Fall 19-		Aug.	Sept.	Oct.	Nov.	Dec.	Jan.	
Sales Percent Each Month								
Sales $	Last Year							
	Plan							
	Revised							
	Actual							
Retail Stock (BOM) $	Last Year							
	Plan							
	Revised							
	Actual							
Markdowns	Last Year							
	Plan (dollars)							
	Last Year Monthly %							
	Plan Monthly %							
	Actual Dollars							
Retail Purchases	Last Year							
	Plan							
	Revised							
	Actual							
Cost $	Last Year							
	Plan							
	Revised							
	Actual							
Ending Stock July 31 Jan 31	Last Year							
	Plan							
	Revised							
	Actual							
Stock to Sales Ratio	Last Year							
	Plan							

Department Buyer _____   Merchandise Manager _____

Controller _____

7. Using the following six-month merchandising plan, fill in last year's figures and then analyze the given information:

Monthly sales percentage each month

Monthly stock-to-sales ratio

Monthly markdown percentage

Monthly planned purchases at retail and cost

Average stock

Stock turnover

Total markdown percentage

Last year's fall season figures were as follows:

	Sales Volume	BOM Stocks	Markdowns
August	$ 95,050	$199,605	$ 6,000
September	$ 98,000	$225,400	$ 9,500
October	$ 87,450	$218,625	$12,500
November	$102,000	$285,600	$ 8,950
December	$195,000	$409,500	$14,000
January	$ 82,000	$220,000	$15,000
February	–	$265,000	–

Average departmental markup was 45%.

8. **Analysis and Interpretation**   This women's robe department, located in Henderson, Wyoming, has completed its second year of fall and holiday business. After analyzing the numbers in Problem 7, discuss the sales flow, the balance of stock, and monthly markdowns. Is the business steady or flat? What suggestions might you offer after you calculate the average stock and the turnover rate?

SIX-MONTH MERCHANDISING PLAN			Department Name:		Plan (This Year)	Actual (Last Year)
			Workroom Cost			
			Cash Discount %			
			Shortage %			
			Average Stock			
Departmental Markup %			Season Stock Turnover			
Last Year	Plan	Actual	Overall Markdown %			

Spring 19-	Feb.	Mar.	Apr.	May	June	July	Season Total
Fall 19-	Aug.	Sept.	Oct.	Nov.	Dec.	Jan.	
Sales Percent Each Month							
Sales $ — Last Year							
Sales $ — Plan							
Sales $ — Revised							
Sales $ — Actual							
Retail Stock (BOM) $ — Last Year							
Retail Stock (BOM) $ — Plan							
Retail Stock (BOM) $ — Revised							
Retail Stock (BOM) $ — Actual							
Markdowns — Last Year							
Markdowns — Plan (dollars)							
Markdowns — Last Year Monthly %							
Markdowns — Plan Monthly %							
Markdowns — Actual Dollars							
Retail Purchases — Last Year							
Retail Purchases — Plan							
Retail Purchases — Revised							
Retail Purchases — Actual							
Cost $ — Last Year							
Cost $ — Plan							
Cost $ — Revised							
Cost $ — Actual							
Ending Stock July 31 Jan 31 — Last Year							
Ending Stock July 31 Jan 31 — Plan							
Ending Stock July 31 Jan 31 — Revised							
Ending Stock July 31 Jan 31 — Actual							
Stock to Sales Ratio — Last Year							
Stock to Sales Ratio — Plan							

Department Buyer _____   Merchandise Manager _____

Controller _____

9. Using the following six-month merchandising plan, fill in last year's figures and then analyze the given information:

   Monthly sales percentage each month

   Monthly stock-to-sales ratio

   Monthly markdown percentage

   Monthly planned purchases at retail and cost

   Average stock

   Stock turnover

   Total markdown percentage

   These figures represent the completion of the first spring season for a new specialty department of home furnishings.

	Sales Volume	Stocks	Markdowns
February	$32,000	$ 57,600	$ 500
March	$24,000	$ 48,000	$1,200
April	$27,000	$ 75,600	$1,500
May	$42,000	$117,600	$5,500
June	$54,000	$108,000	$7,200
July	$36,000	$ 90,000	$4,000
August	–	$102,000	–

   Average departmental markup was 46.5%.

10. **Analysis and Interpretation**   The department in Problem 9 is a first-season home furnishings' shop in a leading department store located in a high-traffic mall and targeting young, well-to-do shoppers. In your opinion, what were the strong points about the first season, after reviewing the sales flow, the balance of stock, markdowns, and turnover rate? Explain. Do you have any suggestions for next year's selling season based on this first-year performance?

		Plan (This Year)	Actual (Last Year)
Department Name:	No.		
Workroom Cost			
Cash Discount %			
Shortage %			
Average Stock			

SIX-MONTH MERCHANDISING PLAN

Departmental Markup %			Season Stock Turnover		
Last Year	Plan	Actual	Overall Markdown %		

Spring 19-		Feb.	Mar.	Apr.	May	June	July	Season Total
Fall 19-		Aug.	Sept.	Oct.	Nov.	Dec.	Jan.	
Sales Percent Each Month								
Sales $	Last Year							
	Plan							
	Revised							
	Actual							
Retail Stock (BOM) $	Last Year							
	Plan							
	Revised							
	Actual							
Markdowns	Last Year							
	Plan (dollars)							
	Last Year Monthly %							
	Plan Monthly %							
	Actual Dollars							
Retail Purchases	Last Year							
	Plan							
	Revised							
	Actual							
Cost $	Last Year							
	Plan							
	Revised							
	Actual							
Ending Stock July 31 Jan 31	Last Year							
	Plan							
	Revised							
	Actual							
Stock to Sales Ratio	Last Year							
	Plan							

Department Buyer _____   Merchandise Manager _____

Controller _____

nine

Changing a Six-Month Plan Using Historical Data

Once you have learned how to analyze numbers and management has made the decision on how to plan for the new season, there are two courses of action you might take. One option is to ignore all of last year's figures and develop a new plan; a second choice is to use last year's figures as the foundation for the future.

The choice to use last year as a foundation can be made when the merchandise managers are comfortable with the overall balance of sales, stocks, markdowns, purchases, and turnover in an area. When a department or operation achieves a good balance, the directives for a new season might be to keep doing things the same way, simply planning for an overall increase. Or, if the economy is struggling, the directive might be to plan for an overall decrease.

When this choice is made, you can simply apply the following math formulas to determine this year's planned figures. What you will be doing is putting a plan together rather than taking it apart. When you analyze, you use division most of the time. When developing a new plan or planning increases or decreases using last year's figures, you use multiplication as your primary tool.

Step 1: Analyze the Given Information

First and foremost, you must analyze the information provided to you on last year's plan. This will give you the basis from which to work. In this case, analyze the following:

Last year's sales percentages by month

Last year's stock-to-sales ratio

Last year's monthly markdown percent

Last year's total markdown percent

Then,

> Calculate last year's monthly purchases at retail and cost.
> Calculate average stock and the stock turnover rate.

Step 2: Determine Sales

After you analyze a plan, it is time to make the changes based on management's directives. Since determining sales is always the first step a buyer takes in analyzing a plan, the same is true when building a new plan. To determine the new sales plan, use these formulas:

Total sales ± percent of increase or decrease = new planned total sales
Season total sales × monthly percent of business = monthly sales

Remember, the sum of the sales percents must be 100%.

Step 3: Figure Out the Stock Data

Once monthly sales have been determined, it is important to have enough stock on hand to support the sales. Here is how you calculate the stock figures for the new sales plan:

> **New monthly planned sales × last year's stock-to-sales ratio (SSR)**
> **= planned BOM**

The monthly SSR was determined when the plan was analyzed.

It is also important to determine the end-of-season stock. By following the formulas given here, you will first compare last year's end-of-season stock to the sales volume for that year and then apply that percentage to determine the new planned EOS dollars.

LY end-of-season stock (EOS) ÷ LY total sales = EOS%
> **TY planned sales × EOS% = TY planned EOS stock**

Step 4: Calculate Markdowns

Calculating markdowns is important to maintain the same promotional pattern that you have presented your consumer in the past. Don't forget to calculate markdowns for both the monthly sales and the overall operation.

To calculate monthly markdown dollars using last year's percentages:

TY monthly sales × monthly % = monthly markdowns

To calculate overall markdown dollars using last year's percentages:

TY planned sales × total markdown % = total markdowns

The monthly markdowns (in dollars) added together must equal the total markdown dollars.

Step 5: Determine Planned Purchases

For planned monthly purchases, calculate purchases at both retail and cost. To evaluate monthly purchases at retail:

**Monthly sales + EOM stock + monthly markdowns
– BOM = purchases**

Another way of calculating how much stock is needed to meet the monthly stock levels is

**BOM – monthly sales – monthly markdowns – EOM
= purchases to restock**

To evaluate monthly purchases at cost:

**100% – departmental MU% = cost%
Purchases × cost% = cost$**

(Don't forget to add the six-months figures at both retail and cost and fill in the totals.)

Step 6: Measure Average Stock and Turnover

You know that the final step in evaluating a department's performance is to measure how the sales and stocks balance by determining the average stock and turnover rate. Using the new sales figures and stock figures, you calculate the average stock and the stock turnover. The stock-turnover rate should be the same as last year's, thereby meeting the management directive to plan the same but implement an increase or decrease based on the economic influences.

To evaluate monthly average stock:

(6 planned BOM + 1 planned EOM) ÷ 7 = planned average stock

To evaluate the stock-turnover rate:

Planned net sales ÷ planned average stock = turnover

■ Problems

1. Using last year's figures for a women's sportswear department, plan a 10% increase for next year, maintaining the same monthly sales percentages, the same stock-to-sales ratios, and the same markdown percents as last year. Last year's spring season figures were as follows:

Sales Volume		Stocks	Markdowns
February	$27,300	$ 83,930	$2,775
March	$50,400	$104,720	$1,109
April	$42,000	$107,030	$5,158
May	$35,700	$ 90,860	$5,200
June	$33,600	$ 90,860	$2,670
July	$21,000	$ 77,000	$1,765
August	–	$ 92,400	–

Last year's departmental markup was 47%. First, fill in last year's information on the six-month plan form and analyze the following statistics:

Monthly sales percentages by month
Last year's stock-to-sales ratio
Last year's monthly markdown percentage
Last year's total markdown percentage

Then,

Calculate last year's monthly purchases at retail and cost.
Calculate average stock and the stock-turnover rate.

Using that information as the foundation, apply the percent of increase to the overall sales and develop a plan using the same percents. Fill in the "Plan" rows.

	Department Name:			No.	
			Plan (This Year)	**Actual** (Last Year)	
SIX-MONTH MERCHANDISING PLAN		Workroom Cost			
		Cash Discount %			
		Shortage %			
		Average Stock			
Departmental Markup %			Season Stock Turnover		
Last Year	Plan	Actual	Overall Markdown %		

Spring 19-			Feb.	Mar.	Apr.	May	June	July	**Season Total**
Fall 19-			Aug.	Sept.	Oct.	Nov.	Dec.	Jan.	
Sales Percent Each Month									
Sales $	Last Year								
	Plan								
	Revised								
	Actual								
Retail Stock (BOM) $	Last Year								
	Plan								
	Revised								
	Actual								
Markdowns	Last Year								
	Plan (dollars)								
	Last Year Monthly %								
	Plan Monthly %								
	Actual Dollars								
Retail Purchases	Last Year								
	Plan								
	Revised								
	Actual								
Cost $	Last Year								
	Plan								
	Revised								
	Actual								
Ending Stock July 31 Jan 31	Last Year								
	Plan								
	Revised								
	Actual								
Stock to Sales Ratio	Last Year								
	Plan								

Department Buyer _____   Merchandise Manager _____

Controller _____

2. Using last year's figures for a jewelry department, plan an 8% increase for next year, maintaining the same monthly sales percentages, the same stock-to-sales ratios and the same markdown percents as last year. Last year's spring season figures were as follows:

	Sales Volume	Stocks	Markdowns
February	$195,000	$462,500	$25,400
March	$ 97,500	$290,000	$ 3,000
April	$102,000	$215,000	$ 3,000
May	$140,000	$282,000	$18,000
June	$106,000	$210,000	$15,000
July	$ 75,500	$165,000	$ 4,000
August	–	$195,000	–

Note: Round your answers to the nearest hundredth to give you the most accurate percentage of sales per month.

Last year's departmental markup was 54%. First, fill in last year's information on the six-month plan form and analyze the following statistics:

Monthly sales percentages by month

Last year's stock-to-sales ratio

Last year's monthly markdown percentage

Last year's total markdown percentage

Then,

Calculate last year's monthly purchases at retail and cost.

Calculate average stock and the stock-turnover rate.

Using that information as the foundation, apply the percent of increase to the overall sales and develop a plan using the same percents. Fill in the "Plan" rows.

SIX-MONTH MERCHANDISING PLAN			Department Name:		No.				
				Plan (This Year)		Actual (Last Year)			
			Workroom Cost						
			Cash Discount %						
			Shortage %						
			Average Stock						
Departmental Markup %			Season Stock Turnover						
Last Year	Plan	Actual	Overall Markdown %						
Spring 19-			Feb.	Mar.	Apr.	May	June	July	Season Total
Fall 19-			Aug.	Sept.	Oct.	Nov.	Dec.	Jan.	
Sales Percent Each Month									
Sales $		Last Year							
		Plan							
		Revised							
		Actual							
Retail Stock (BOM) $		Last Year							
		Plan							
		Revised							
		Actual							
Markdowns		Last Year							
		Plan (dollars)							
		Last Year Monthly %							
		Plan Monthly %							
		Actual Dollars							
Retail Purchases		Last Year							
		Plan							
		Revised							
		Actual							
Cost $		Last Year							
		Plan							
		Revised							
		Actual							
Ending Stock July 31 Jan 31		Last Year							
		Plan							
		Revised							
		Actual							
Stock to Sales Ratio		Last Year							
		Plan							

Department Buyer _____   Merchandise Manager _____

Controller _____

3. Using last year's figures for a luggage department, plan a 10% decrease for this next year, maintaining the same monthly sales percentages, the same stock-to-sales ratios, and the same markdown percents as last year. Last year's fall season figures were as follows:

	Sales Volume	BOM Stocks	Markdowns
August	$39,600	$130,680	$1,650
September	$42,400	$132,990	$2,640
October	$46,200	$133,980	$5,610
November	$66,000	$165,000	$3,960
December	$72,100	$159,720	$9,900
January	$62,700	$163,020	$9,240
February	–	$150,000	–

Last year's departmental markup was 42%. First, fill in last year's information on the six-month plan form and analyze the following statistics:

Monthly sales percentages by month

Last year's stock-to-sales ratio

Last year's monthly markdown percentage

Last year's total markdown percentage

Then,

Calculate last year's monthly purchases.

Calculate average stock and the stock-turnover rate.

Using that information as the foundation, apply the percent of decrease to the overall sales and develop a plan with the same percents you already analyzed. Fill in the "Plan" rows.

SIX-MONTH MERCHANDISING PLAN	Department Name:			No.		
			Plan (This Year)		**Actual** (Last Year)	
	Workroom Cost					
	Cash Discount %					
	Shortage %					
	Average Stock					
Departmental Markup %	Season Stock Turnover					
Last Year	Plan	Actual	Overall Markdown %			

Spring 19-		Feb.	Mar.	Apr.	May	June	July	**Season Total**
Fall 19-		Aug.	Sept.	Oct.	Nov.	Dec.	Jan.	
Sales Percent Each Month								
Sales $	Last Year							
	Plan							
	Revised							
	Actual							
Retail Stock (BOM) $	Last Year							
	Plan							
	Revised							
	Actual							
Markdowns	Last Year							
	Plan (dollars)							
	Last Year Monthly %							
	Plan Monthly %							
	Actual Dollars							
Retail Purchases	Last Year							
	Plan							
	Revised							
	Actual							
Cost $	Last Year							
	Plan							
	Revised							
	Actual							
Ending Stock July 31 Jan 31	Last Year							
	Plan							
	Revised							
	Actual							
Stock to Sales Ratio	Last Year							
	Plan							

Department Buyer _____   Merchandise Manager _____

Controller _____

4. Using last year's figures for a garden center, plan a 15% increase for this next year, maintaining the same monthly sales percentages, the same stock-to-sales ratios, and the same markdown percents as last year. Last year's fall season figures were as follows:

	Sales Volume	BOM Stocks	Markdowns
February	$16,500	$41,250	$1,000
March	$24,750	$44,550	$1,500
April	$33,000	$60,000	$1,000
May	$35,750	$58,000	$4,000
June	$33,000	$57,000	$2,000
July	$22,000	$40,000	$1,800
August	–	$42,000	–

Last year's departmental markup was 40%. First, fill in last year's information on the six-month plan form and analyze the following statistics:

Monthly sales percentages by month

Last year's stock-to-sales ratio

Last year's monthly markdown percentage

Last year's total markdown percentage

Then,

Calculate last year's monthly purchases at retail and cost.

Calculate average stock and the stock-turnover rate.

Using that information as the foundation, apply the percent of increase to the overall sales and develop a plan with the same percents. Fill in the "Plan" rows.

		Plan (This Year)	Actual (Last Year)
	Department Name:		No.

SIX-MONTH MERCHANDISING PLAN		Plan (This Year)	Actual (Last Year)
	Workroom Cost		
	Cash Discount %		
	Shortage %		
	Average Stock		

Departmental Markup %			Season Stock Turnover	
Last Year	Plan	Actual	Overall Markdown %	

Spring 19-			Feb.	Mar.	Apr.	May	June	July	Season Total
Fall 19-			Aug.	Sept.	Oct.	Nov.	Dec.	Jan.	
Sales Percent Each Month									
Sales $		Last Year							
		Plan							
		Revised							
		Actual							
Retail Stock (BOM) $		Last Year							
		Plan							
		Revised							
		Actual							
Markdowns		Last Year							
		Plan (dollars)							
		Last Year Monthly %							
		Plan Monthly %							
		Actual Dollars							
Retail Purchases		Last Year							
		Plan							
		Revised							
		Actual							
Cost $		Last Year							
		Plan							
		Revised							
		Actual							
Ending Stock July 31 Jan 31		Last Year							
		Plan							
		Revised							
		Actual							
Stock to Sales Ratio		Last Year							
		Plan							

Department Buyer _____   Merchandise Manager _____

Controller _____

5. Using last year's figures for a stereo entertainment department, plan a 5% decrease for this next year, maintaining the same monthly sales percents, the same stock-to-sales ratios, and the same markdown percents as last year. Last year's fall season figures were as follows:

	Sales Volume	BOM Stocks	Markdowns
August	$ 68,250	$128,475	$ 1,000
September	$126,000	$190,675	$ 1,500
October	$105,000	$198,250	$20,000
November	$ 89,250	$157,150	$ 2,500
December	$ 84,000	$157,150	$ 7,500
January	$ 52,500	$122,500	$17,000
February	–	$161,000	–

Last year's departmental markup was 42%. First, fill in last year's information on the six-month plan form and analyze the following statistics:

Monthly sales percentages by month
Last year's stock-to-sales ratio
Last year's monthly markdown percentage
Last year's total markdown percentage

Then,

Calculate last year's monthly purchases at retail and cost.
Calculate average stock and the stock turnover rate.

Then using that information as the foundation, apply the percentage of decrease to the overall sales and develop a plan with the same percentages. Fill in the "Plan" rows.

			Plan (This Year)		**Actual** (Last Year)	
		Department Name:		No.		

SIX-MONTH MERCHANDISING PLAN		Workroom Cost		
		Cash Discount %		
		Shortage %		
		Average Stock		

Departmental Markup %			Season Stock Turnover		
Last Year	Plan	Actual	Overall Markdown %		

		Feb.	Mar.	Apr.	May	June	July	**Season Total**
Spring 19-								
Fall 19-		Aug.	Sept.	Oct.	Nov.	Dec.	Jan.	
Sales Percent Each Month								
Sales $	Last Year							
	Plan							
	Revised							
	Actual							
Retail Stock (BOM) $	Last Year							
	Plan							
	Revised							
	Actual							
Markdowns	Last Year							
	Plan (dollars)							
	Last Year Monthly %							
	Plan Monthly %							
	Actual Dollars							
Retail Purchases	Last Year							
	Plan							
	Revised							
	Actual							
Cost $	Last Year							
	Plan							
	Revised							
	Actual							
Ending Stock July 31 Jan 31	Last Year							
	Plan							
	Revised							
	Actual							
Stock to Sales Ratio	Last Year							
	Plan							

Department Buyer _____   Merchandise Manager _____

Controller _____

ten

Developing New Six-Month Plans

Often a company will tell a buyer to forget the past and develop a six-month plan based on new goals. Sometimes you do not have an old department to refer to, or you need a new plan because you are opening a new store or department. The situations may be different, but the procedures are the same and the focus is very simple.

Once you have completed your research involving trends, customers, pricing, and merchandise, putting together the financial plan will be easy. Don't be afraid to *change, change,* and *change* all the figures so you can effectively develop a plan for success. Rarely does a buyer come up with a perfect plan in the first couple of tries. Sometimes you won't think the turnover is effective enough, or you won't like the markdown structure. Remember what this is about—developing a new plan.

Merchandise planning is actually fun. It is like putting all the pieces of a puzzle together to create a picture or even like reading a good book. When you read a book, each chapter must keep you focused and excited so you will keep on reading. This is true for a store, too. Each month is like a chapter. It is hoped that you will inspire and attract your audience to want more and more, but even as your book must draw to a close, so does your season. The trick is to make it a best-seller so your customer will shop into the next season, just as you will buy a second book.

As you probably have already realized, since you are building a plan, most of your math decisions are determined using multiplication.

Step 1: Analyze the Given Information

Discuss the trends of the economy, the industry, the market, and the consumer. Look at the forecasting projections of the customers' wants, needs, and lifestyles. Identify key marketing tools that will encourage a strong consumer response based on those needs. Without research, the numbers will be meaningless. But, with solid research, focused financial plans can easily be developed, and exciting marketing plans can be set in motion.

Step 2: Determine Sales

- Determine your planned sales for the 6 months.
- Determine the percentage of business that you feel can be generated each month.
- Remember, all the percentages must add up to 100%.

Once the sales volume for 6 months has been forecast and the monthly percentages have been established, use the following formula to determine the monthly sales plan:

Monthly sales = monthly % × total sales volume planned

Step 3: Figure Out the Stock Data

- Determine the stock-to-sales ratio (SSR).
- Calculate the beginning-of-the-month (BOM) stocks.

Plan a stock-to-sales ratio each month that you feel will provide a good mix of merchandise for the department. Remember to increase stocks prior to busy seasons so there is a good merchandise assortment. Again, there is no formula for calculating this number; it is based on how much stock is needed at the beginning of a period to support the sales in that same period. Once the SSR has been established, using the following formula you can determine the monthly BOM stock:

BOM stock = monthly sales × monthly stock-to-sales ratio

After the monthly stocks have been determined, a closing stock figure has to be established.

- Determine the EOS stock.

Again, there are no formulas to use in calculating this figure. Instead, review the BOM plan for the 6 months and determine a stock figure that will be low enough to clean out the old season but strong enough to welcome the new. With the end-of-season stock determined, you can calculate both the average stock figure and the turnover rate.

At this point, before you go into markdowns or purchases, it is important to measure how the plan is developing. This is done by checking the stock-turnover rate. If the stock turnover is too sluggish or, conversely, too fast, this is the time to go back and modify the stock-to-sales ratios before moving ahead. To evaluate the turnover (T/O) rate:

First, calculate the average stock:
$$(6 \text{ BOM} + 1 \text{EOS}) \div 7 = \text{average stock}$$

Then, calculate the turnover rate:
$$\text{Net sales} \div \text{average stock} = \text{T/O}$$

See if you are moving your department in the right direction. *Remember:* Check the FOR, prepared by the National Retail Federation, for guidance.

Step 4: Calculate Markdowns

At this point it is time to determine the total markdowns. Check the FOR to guide you on the industry's general performance for a similar type of store or department. You will find the data will give you a range of percents from low to high of what similar stores are generating in markdowns. Once you determine the overall markdown percentage for the 6 months, use the following formula to determine the overall markdown dollars:

Total markdown dollars = net sales × overall markdown%

Next, plan the monthly markdowns by allocating the total markdown dollars over a 6-month period. Just split up the markdown dollar allowance according to your marketing plans for advertising and promotion. For example, if you have $5,000 in total season markdowns and your company plans an anniversary sale, you may find yourself using $3,000 of the $5,000 in markdown money in one month. That would leave only $2,000 to disburse among the other 5 months. You may put most of it into another month to run another clearance sale and allocate only a few hundred dollars to the remaining 4 months just to cover employee discounts and customer allowances.

To determine the monthly markdown percent:

Monthly markdown% = monthly markdowns ÷ monthly sales

Step 5: Plan Monthly Purchases

It is important to determine the planned purchases at both cost and retail to visualize the flow of merchandise each month. Use a planned departmental markup that is within the industry averages. Again, this information can be found in both your research and by consulting the using industry standards prepared by the NRF.

Calculate the purchases with the same formulas used in analyzing plans:

Monthly sales + EOM + monthly markdowns – BOM
= monthly purchases at retail

Then, working with the monthly retail purchases, it is very important to figure the cost:

100% – markup% = cost%
Retail purchases × cost% = cost purchases

■ Problems

1. A major skate company is expanding its business by opening skate stores throughout the United States. The product line consists of equipment (for in-line skating, ice skating, and hockey), clothing, instructional services, minicamps and in-house promotions. Based on the success of stores along the east coast, this manufacturer plans to open a store for the upcoming spring season in Houston, Texas.

Financial projections are as follows:

- Planned sales are $700,000 for the spring season.
- Average departmental MU is planned at 52%.
- Planned markdowns are 12% of planned sales.
- The first and last months of the 6 months show weak sales trends.
- A stock assortment with a 3.0 to 1 stock-to-sales ratio is necessary to build for opening.
- Management will accept an EOS stock figure of $220K.

Based on these data, prepare a six-month plan. Upon completion of the financial plans, prepare to discuss professionally an analysis of these financial decisions.

1. Analysis of the monthly sales projections.
2. Analysis of the monthly SSR and monthly BOM stock.
3. Analysis of why average stock is substantial.
4. Determination of why the turnover is in line with industry standards and with the opening of a new store.
5. Explanation of the breakdown of markdowns in relation to the company's marketing and promotional plans.
6. Analysis of the flow of new merchandise monthly.
7. Overview of monthly marketing, sales, and promotional plans that correlate with the sales projections.

Remember, your financial budget must be in line with consumer, economic, and fashion trends, along with the trends of the geographic area being targeted for the new store.

Note: Marketing consultants are responsible for providing psychographic and demographic data for business owners to use in making sound decisions. The information would be in the form of a detailed marketing research report.

			Department Name:		No.			
					Plan (This Year)	**Actual** (Last Year)		
SIX-MONTH MERCHANDISING PLAN			**Workroom Cost**					
			Cash Discount %					
			Shortage %					
			Average Stock					
Departmental Markup %			**Season Stock Turnover**					
Last Year	Plan	Actual	**Overall Markdown %**					

Spring 19- Fall 19-			Feb. Aug.	Mar. Sept.	Apr. Oct.	May Nov.	June Dec.	July Jan.	**Season Total**
Sales Percent Each Month									
Sales $		Last Year							
		Plan							
		Revised							
		Actual							
Retail Stock (BOM) $		Last Year							
		Plan							
		Revised							
		Actual							
Markdowns		Last Year							
		Plan (dollars)							
		Last Year Monthly %							
		Plan Monthly %							
		Actual Dollars							
Retail Purchases		Last Year							
		Plan							
		Revised							
		Actual							
Cost $		Last Year							
		Plan							
		Revised							
		Actual							
Ending Stock July 31 Jan 31		Last Year							
		Plan							
		Revised							
		Actual							
Stock to Sales Ratio		Last Year							
		Plan							

Department Buyer _____   Merchandise Manager _____

Controller _____

2. A children's toy department is being set up in a major discount store in Nebraska for the fall season. Although the company recognizes that the majority of sales will come during the holiday selling season, they also know that they have to bring merchandise in as early as August 1 to attract customer interest. This department is being considered a test market. If the response is in line with what the analysts have projected, the discount store will consider funding money and floor space for this department year-round.

Based on the studies, the following are being projected:

- Planned sales are $600,000 for the fall season.
- Planned markdowns are 10% of planned sales, with the majority of the markdowns falling in January.
- The slowest sales are in the first and last months of the 6 months, and a stock assortment of about 3.0 to 1 stock-to-sales ratio is necessary for opening.
- A closing stock figure of no more than $250,000 is being asked for by management. $250K would be large enough to cover a moderate stock assortment but low enough to have cleared out holiday items successfully.
- The planned departmental markup goal is 50.5%.

Upon completion of the financial plans, discuss professionally an analysis of these financial decisions:

1. Analysis of the monthly sales projections.
2. Analysis of the monthly SSR and monthly BOM stock.
3. Analysis of why average stock is substantial.
4. Determination of why the turnover is in line with industry standards and with the opening of a new store.
5. Explanation of the breakdown of markdowns in relation to the company's marketing and promotional plans.
6. Analysis of the flow of new merchandise monthly.
7. Overview of monthly marketing, sales, and promotional plans that correlate with the sales projections.

SIX-MONTH MERCHANDISING PLAN			Department Name:		No.	
				Plan (This Year)	Actual (Last Year)	
			Workroom Cost			
			Cash Discount %			
			Shortage %			
			Average Stock			
Departmental Markup %			Season Stock Turnover			
Last Year	Plan	Actual	Overall Markdown %			

Spring 19-	Feb.	Mar.	Apr.	May	June	July	Season Total
Fall 19-	Aug.	Sept.	Oct.	Nov.	Dec.	Jan.	
Sales Percent Each Month							
Sales $ Last Year							
Plan							
Revised							
Actual							
Retail Stock (BOM) $ Last Year							
Plan							
Revised							
Actual							
Markdowns Last Year							
Plan (dollars)							
Last Year Monthly %							
Plan Monthly %							
Actual Dollars							
Retail Purchases Last Year							
Plan							
Revised							
Actual							
Cost $ Last Year							
Plan							
Revised							
Actual							
Ending Stock July 31 Jan 31 Last Year							
Plan							
Revised							
Actual							
Stock to Sales Ratio Last Year							
Plan							

Department Buyer _____   Merchandise Manager _____

Controller _____

3. A store selling reproductions of antiques is opening on a major tourist road in Chagrin Falls, Ohio. This area attracts shoppers from a 100-mile radius with its quaint gift stores and home-crafted items. Several antique stores have done well in this trading area, and a young entrepreneur plans to sell antique reproductions in a 1,500-square-foot store.

Based on the research, the owner is planning an opening season in August when traffic begins to increase. The financial projections are as follows:

- Planned sales of $175,000, with November and December as the prime selling months
- A beginning stock-to-sales ratio of 3.0 to 1
- Markdowns of only 6%
- Closing stock on January 31 of $75,000
- Planned departmental markup of 50%

Upon completion of the six-month plan, develop a marketing plan that would support the flow of sales and stock reductions, and describe how it will be important to the development of this new store.

SIX-MONTH MERCHANDISING PLAN		Department Name:		No.	
				Plan (This Year)	**Actual** (Last Year)
		Workroom Cost			
		Cash Discount %			
		Shortage %			
		Average Stock			
Departmental Markup %		**Season Stock Turnover**			
Last Year	Plan	Actual	**Overall Markdown %**		

Spring 19-		Feb.	Mar.	Apr.	May	June	July	**Season Total**
Fall 19-		Aug.	Sept.	Oct.	Nov.	Dec.	Jan.	
Sales Percent Each Month								
Sales $	Last Year							
	Plan							
	Revised							
	Actual							
Retail Stock (BOM) $	Last Year							
	Plan							
	Revised							
	Actual							
Markdowns	Last Year							
	Plan (dollars)							
	Last Year Monthly %							
	Plan Monthly %							
	Actual Dollars							
Retail Purchases	Last Year							
	Plan							
	Revised							
	Actual							
Cost $	Last Year							
	Plan							
	Revised							
	Actual							
Ending Stock July 31 Jan 31	Last Year							
	Plan							
	Revised							
	Actual							
Stock to Sales Ratio	Last Year							
	Plan							

Department Buyer _____   Merchandise Manager _____

Controller _____

4. A boutique gift shop is opening on a cruise ship that tours Alaska for the spring traveling season. The shop is to showcase items unique to the Alaskan cultures. Included in the stock assortment will be gifts, stationery, books, tapes, maps, and pottery items. The only customers will be ship's passengers. Stocks need to be exciting and unique but with a very tight stock-to-sales ratio. The company has researched other gift shops and knows that there is strong potential; however, their plan is to start off slowly for the spring season.

- Planned sales of $80,000
- A beginning stock-to-sales ratio of 2.5 to 1
- Markdowns of only 5%
- Closing stock to be planned low at this time but to be modified later on if this new venture gathers strong consumer response
- Planned departmental markup of 48%

Upon completion of the six-month plan, develop a marketing plan that would support the flow of sales and stock reductions, and explain how it will be important to the development of this boutique operation.

			Department Name:			No.			
					Plan (This Year)		**Actual** (Last Year)		
SIX-MONTH MERCHANDISING PLAN			Workroom Cost						
			Cash Discount %						
			Shortage %						
			Average Stock						
Departmental Markup %			Season Stock Turnover						
Last Year	Plan	Actual	Overall Markdown %						

Spring 19-			Feb.	Mar.	Apr.	May	June	July	**Season Total**
Fall 19-			Aug.	Sept.	Oct.	Nov.	Dec.	Jan.	
Sales Percent Each Month									
Sales $		Last Year							
		Plan							
		Revised							
		Actual							
Retail Stock (BOM) $		Last Year							
		Plan							
		Revised							
		Actual							
Markdowns		Last Year							
		Plan (dollars)							
		Last Year Monthly %							
		Plan Monthly %							
		Actual Dollars							
Retail Purchases		Last Year							
		Plan							
		Revised							
		Actual							
Cost $		Last Year							
		Plan							
		Revised							
		Actual							
Ending Stock July 31 Jan 31		Last Year							
		Plan							
		Revised							
		Actual							
Stock to Sales Ratio		Last Year							
		Plan							

Department Buyer _____   Merchandise Manager _____

Controller _____

5. A major department store is expanding its business by opening linen stores in several of the major malls in the United States. Major competition will come from stores such as Linens and Things, Linen Supermarket, and the Home Stores. Due to the success of these major competitors, this leading department store on the west coast of the United States plans on opening a prototype store in the spring season. Based on extensive research of the market and the competition, the following financial goals are being set:

- Planned sales are $900,000 in the first season
- Average departmental MU is planned at 45%
- Planned markdowns are 12% of planned sales
- The last 2 months of the season show weak sales trends
- You have been asked to establish a stock-to-sales ratio each month that will produce a minimum 2.0 stock-to-sales ratio for the season

Based on these data, prepare a six-month plan. Upon completion of the financial plans, prepare to discuss professionally an analysis of these financial decisions:

1. Monthly sales projections
2. Monthly SSR and monthly BOM stock
3. Why average stock is substantial
4. Why the turnover is in line with industry standards and with the opening of a new store
5. The breakdown of markdowns in relation to the company's marketing and promotional plans
6. The flow of new merchandise monthly
7. Overview of monthly marketing, sales, and promotional plans that correlate with the sales projections

(Remember, your financial budget must be in line with consumer, economic, and fashion trends, along with the trends of the geographic area being targeted for the new store.

Note: Marketing consultants are responsible for providing psychographic and demographic data for business owners to use in making sound decisions. The information would be in the form of a detailed marketing research report.)

SIX-MONTH MERCHANDISING PLAN	Department Name:		No.	
			Plan (This Year)	**Actual** (Last Year)
	Workroom Cost			
	Cash Discount %			
	Shortage %			
	Average Stock			

Departmental Markup %			Season Stock Turnover		
Last Year	Plan	Actual	**Overall Markdown %**		

Spring 19-			Feb.	Mar.	Apr.	May	June	July	**Season Total**
Fall 19-			Aug.	Sept.	Oct.	Nov.	Dec.	Jan.	
Sales Percent Each Month									
Sales $		Last Year							
		Plan							
		Revised							
		Actual							
Retail Stock (BOM) $		Last Year							
		Plan							
		Revised							
		Actual							
Markdowns		Last Year							
		Plan (dollars)							
		Last Year Monthly %							
		Plan Monthly %							
		Actual Dollars							
Retail Purchases		Last Year							
		Plan							
		Revised							
		Actual							
Cost $		Last Year							
		Plan							
		Revised							
		Actual							
Ending Stock July 31 Jan 31		Last Year							
		Plan							
		Revised							
		Actual							
Stock to Sales Ratio		Last Year							
		Plan							

Department Buyer _____   Merchandise Manager _____

Controller _____

Calculating Open to Buy

Open to Buy

Open-to-buy reports and calculations are done by merchandisers monthly to make sure enough merchandise has been purchased to meet the planned stocks calculated on the six-month plan. The OTB report is a key control tool for a buyer, one that allows a business owner to keep the merchandise flowing into a department. It serves as a guide for the merchant to determine how much to purchase if a department needs more merchandise, based on what the buyer planned to purchase on the six-month merchandising plan, and also quickly shows a buyer if a department is getting too much merchandise in reserve. Too much merchandise is as much of a problem as too little. By calculating the open to buy (OTB), a merchandiser can establish controls which reflect good merchandise timing.

Industry **Terms** and *jargon*

OTB   Open to buy, the amount of dollars available to spend during a specific period of time. OTB is a phrase that refers to the amount of retail dollars that have been allocated to be purchased based on what is needed, as determined by the monthly sales and what already has been ordered for the month. Sometimes abbreviated as O.T.B. (Review Chapter 5.)

OO   On order, the value of merchandise that has been ordered but not yet received. (Review Chapter 5.)

Overbought   A condition that occurs when a buyer has too much merchandise on order and scheduled for delivery in a particular month. When a buyer is overbought, they cannot take advantage of any vendor special purchases, and if the overbought condition overlaps into other months, it can also slow down the turnover and increase markdowns.

Purchase Journal A financial recap report identifying merchandise that has been ordered. Tracked daily by computer, this report calculates merchandise receipts and outstanding orders. This financial tool is important for merchants to use in maintaining a strong merchandise flow.

Determining Open to Buy

The dollars available for merchandise needed are calculated as follows:

> **Planned purchases for the month (found on six-month plan)**
>
> **– Orders already placed and expected for delivery that month**
>
> **= Open to buy for the month**

To do this calculation, you will need the planned purchases determined on the six-month plan. Let's consider an example.

EXAMPLE

Planned sales for July are $220,000. The BOM stock is planned at $315,000. Orders already placed for July are $172,000. The EOM stock is $300,000. What is the July OTB?

	$220,000	(Planned sales)
+	$300,000	(EOM stock)
–	$315,000	(BOM stock)
=	$205,000	(Planned purchases)

Now to determine the open to buy:

	$205,000	(Planned purchases)
–	$172,000	(Purchases already on order)
=	$ 33,000	OTB

In Chapter 5 you learned how to balance stocks by averaging prices of merchandise on a buying plan. The open-to-buy report is another way to help you plan purchases by cross-checking stock flow with sales forecasts to maintain ideal stock assortments for the consumer.

This information is generally computer-generated and is followed weekly by buyers. It can be used to determine the stock on hand, to verify information about the receipts indicated on the purchase journal, and to assist a buyer in making any adjustments based on sales and markdowns needed to be in line with the six-month planned BOM stock projections.

■ Problems

1. In August the BOM stock is $32,500. Planned sales are $18,600. During the first week, $6,200 worth of merchandise comes in, and the planned stock for September is $21,000. What is the OTB for August?

2. Planned sales for April are $100,000. The BOM stock is planned at $185,000. Orders placed for April come to $72,000. The May BOM is $215,000. What is the April OTB?

3. On November 1 the BOM stock is $95,000, and the EOM is $136,000. The purchases received in the first week totaled $5,750, with an extra $22,000 still on order. The sales plan for the month is $26,780. What is the OTB?

4.

September BOM	$360,000
September EOM	$440,000
September sales	$148,000
September markdowns	$ 23,500
September 1–9 purchases received	$ 42,000
September outstanding orders	$106,000

Determine the September OTB. _____

(*Hint*: Remember how you calculated planned purchases on the six-month plan. Begin there and then finish the problem by determining the OTB.)

5. An open-to-buy report was prepared on January 15. The following information was provided:

January BOM	$ 86,000
January EOM	$104,000
January sales	$ 52,000
January markdowns	$ 10,500
January 5–12 purchases received	$ 11,000
January outstanding orders	$ 6,000
Determine the January OTB.	_____

With the amount of merchandise on order, and noting that this open to buy has been prepared on the 15th, what concerns might a buyer face after reviewing these figures?

6. An open-to-buy report was prepared on August 5 for a linen department. The following information was provided:

August BOM	$316,000
August EOM	$325,000
August sales	$152,000
August markdowns	$ 31,450
August 1–5 purchases received	$ 46,000
August outstanding orders	$187,000
Determine August OTB.	_____

The amount of merchandise on order indicates that the buyer will be over the planned BOM stock plan for September 1. What are some steps a buyer can take to resolve this situation and still stay in line with the original merchandising plan?

7. The retail inventory for men's ties in the men's accessories department of Wilson's Department Store was $62,000 on May 1, with a planned EOM figure of $56,000. The planned sales are $44,000 for the month, markdowns are $800, and orders placed already totaled $8,000 at retail. What is the open to buy?

8. The Box Stop Shop, which sells unique gift items, plans to open a new division on November 1 with $85,000 of stock. Planned sales are $56,000, with markdowns at $1,000 only and an ending stock figure of $103,000. As of October 14, $36,000 of mer-

chandise is on order. As the owner prepares for a trip to market, what is the OTB for November?

9. Chrissy's Campus started August 1 at $65,000. This amount was $8,000 over the planned BOM stock figure. When the merchandise manager reviewed the purchase journal, he told the buyer to cancel orders to avoid being overbought. Look at the following figures and determine how much Chrissy's Campus would be overbought if the buyer does not cancel any orders.

Planned BOM	=	$57,000
Actual BOM	=	$65,000
Planned sales	=	$32,000
Planned markdowns	=	$ 6,500
Planned EOM	=	$48,000
Purchases on order	=	$31,500

Would the store have been overbought if the operation had come in on the planned BOM figure? What would the figures be? Based on these figures, what are some areas that you would research to understand the stock coming in over plan?

10. One of the biggest concerns a buyer has is timing the merchandise flow into a department. If merchandise is on order but arrives late in the month, several days of selling opportunities may be missed. If the merchandise comes in all at one time, stocks may be too heavy and not enough will be sold. That merchandise becomes old and can result in a markdown. Looking at the following numbers, determine the open to buy, and discuss any concerns you might have. Also, based on today's date, February 25, discuss any ideas you might have to stimulate earlier shipments or overcome late-arriving merchandise. Be creative. Think how you could share the new merchandise through promotional activities for your consumer.

Today's date:	February 25
March BOM	$185,000
March 1–5 on order	$ 5,600
March 6–12 on order	$ 10,800
March planned sales	$ 72,000
March 15–20 on order	$ 23,000
March markdowns	$ 8,050
April BOM	$215,000
Open to buy	_____

twelve

Classification Planning

Classification planning sorts out the inventory identified on the six-month plan. As a merchandiser you calculate only the total stock dollars on a six-month plan. Now, you have to build the "perfect" stock for your customers. Building a balance does not happen simply because of a keen awareness of the trends; instead it is a planned-out evaluation of what merchandise is needed to make the desired sales. Classification planning helps a buyer focus on merchandise categories, prices, styles, sizes, and even colors. It is the best way to mix together an ideal stock assortment, because a buyer can control the balance and use the model stock plan with the open-to-buy dollars and still maintain a strong inventory control.

Industry **Terms** and *jargon*

Classification Plans   A report or plan that groups similar types of merchandise together by price points.

Model Stock   The "perfect" amount of merchandise on hand to meet customer demands and needs. It occurs when stock is broken down by styles and competitive price points to provide the best assortment to the consumer. This type of planning is also called *unit planning* or *assortment planning* because it allows a merchandiser to specifically identify how many pieces of merchandise are available in a category of merchandise at a specific price point and in a general style and color range, if applicable.

Assortment Plans   Another term for classification plans. Stock or inventory plans that mix merchandise to create an exciting inventory. Sometimes this is referred to as *unit planning* or *model stock planning*.

Merchandise Mix   The blending of product to provide the consumer with choices.

Price Range   Merchandise prices from a low to a high point in a specific department or for a specific manufacturer.

Price Line   Sometimes called *price points*, specific prices that customers are willing to pay for a variety of merchandise in a specific department.

Price Points   Key prices for certain types of merchandise in a department.

Units   Pieces.

SKU   Stock keeping units.

The easiest way to explain classification and model stock planning is through an example. You will see that a well-developed model stock takes the following into consideration:

- Six-month plan reflecting previous departmental historical data
- Open to buy (OTB)
- Fashion trends and consumer lifestyles
- Department price range and current economic and sales trends

The following example will take you through the planning process step by step.

Building a Model Stock Through Classification Planning

To plan a model stock, you will review and then plan the assortment in the following order:

1. Identify the department's past sales records.
2. Focus on the most salable price points.
3. Identify the styles that are most popular in this area for this type of customer.
4. Identify the fashion colors.
5. Identify the key sizes.
6. Break down the specific number of pieces.

As each one of these areas is addressed, you use a classification chart to guide you through the development and planning. Forms often vary throughout the industry, because they are targeted to specific departmental needs. Let's look at a typical chart and see how it can be a useful tool for a buyer in planning.

Elements of a Typical Classification Plan

A The name of the classification being developed. Classifications also have an identifying classification number, which is used in the plan.

B The season or month being developed. Sometimes buyers will work month by month; sometimes the classifications plans are developed for the season.

C The amount of retail stock that is in the department for either the season or month, whatever is being addressed in B.

D The percent that the classification is of the total amount.

E The dollar value determined for the classification. Yes, those helpful hints are back again:

Total amount × percent = specific amount

F Price ranges. Most operations address the price range and can identify the low, middle, and high price points in a classification of merchandise. Once those key price points are identified, the buyer can review past records and industry trends to find out what percentage of business was generated in those areas.

G The percentage values that the buyer has determined.

H The total retail dollars by price point, determined by calculating:

Total planned dollars × percent = specific retail for that price point

Helpful Hint #1

I Once the total retail per classification is determined, the quotient of the total divided by the price point. From this, the buyer will know how many units are available per price point.

J A further breakdown by styles, sizes, and then colors.

A Specific Example

Let's work a problem out on a chart so you can see how to break down thousands of dollars into just a handful of pieces.

The goal is to develop the casual men's sweaters, classification 8, which has $30,000 of retail stock on November 1. This dollar amount was determined after the buyer reviewed the historical sales data for this classification and combined that information with the consumer, economic, and fashion trends to determine what percentage and what dollar value should be allocated to the classification as part of the total BOM stock.

Classification Planning Chart

Classification _____A_____ (Name and identifying number)

Season or month _____B_____ Total BOM stock _____C_____

Classification % to total stock _____D_____ Total planned dollars _____E_____

	Price Line	% Value	Total Retail $	Total Units
Lower	F	G	H	I
Middle	_____	_____	_____	_____
Upper	_____	_____	_____	_____

Inventory Breakdown

Lower	Middle	Upper
Unit _____Pieces/units_____	Units _____Pieces/units_____	Units _____Pieces/units_____

Lower	J	**Middle**		**Upper**	
Style	Style	Style	Style	Style	Style
_____	_____	_____	_____	_____	_____
%	%	%	%	%	%
Pieces/units	Pieces/units	Pieces/units	Pieces/units	Pieces/units	Pieces/units
Subclassification		Subclassification		Subclassification	

	Style	Style	Style	Style	Style	Style	Style	Style	Style	Style	Style	Style
	%	%	%	%	%	%	%	%	%	%	%	%
	Pcs	Pcs	Pcs	Pcs	Pcs	Pcs	Pcs	Pcs	Pcs	Pcs	Pcs	Pcs
Small	___	___	___	___	___	___	___	___	___	___	___	___
Med.	___	___	___	___	___	___	___	___	___	___	___	___
Large	___	___	___	___	___	___	___	___	___	___	___	___
X-Large	___	___	___	___	___	___	___	___	___	___	___	___

The buyer determined that 12% of the monthly sales would come from casual men's sweaters. The monthly stock is planned at $250,000, so 12% of that is the $30,000 needed for the casual sweater classification. Keep in mind that the buyer knows this inventory will be profitable only if the sweaters appeal to the customer and a good variety of sizes and colors is available.

Helpful Hint #1

$250,000 × 12% = $30,000

The following step-by-step planning was done for the $30,000 stock of men's sweaters for November 1, BOM stock.

Step 1: Price Points   The buyer must determine the key price lines for sweaters in this department. After a review, the buyer determined that the key points are as follows:

Lower price line is $20.00.
Middle price line is $28.00.
High/upper price line is $36.00.

We enter these amounts on the chart on page 199.

Step 2: Sales Allocations   After reviewing past sales, the buyer determines that the lower price line makes up 20% of the sweater sales, the middle price line represents 50% of sales, and the remaining 30% fall into the upper price line of $36.00.

We again enter these amounts on the chart (page 199) and calculate the retail value for each price line.

20% of $30,000 = $6,000

50% of $30,000 = $15,000

30% of $30,000 = $9,000

Helpful Hint #1

Step 3   Pieces or units per price line   Once the buyer has determined the dollar value for each line, it is time to determine styles, sizes, and colors. This is done most efficiently by working with pieces. Therefore, divide the dollar value in each price line by the unit price. This will tell us by price line how many units we have to work with.

Lower price: $6,000 ÷ $20.00   =   _300_ pieces, or units

Middle price: $15,000 ÷ $28.00   =   _536_ pieces, or units

High-end price: $9,000 ÷ $36.00   =   _250_ pieces, or units

We enter these amounts on page 199.

Step 4 Styling Within each price line the buyer plans to carry two basic styles: cardigans and pullovers. The fashion trends indicate that pullovers are far more popular in the low-end price. Pullovers are a little more popular in the middle price range and in the high end, it is about a 50–50 split.

Based on this information, we determine the following breakdown.

Lower price: Pullovers 75% of stock, cardigans 25% of stock
Middle price: Pullovers 60% of stock, cardigans 40% of stock
High-end price: Pullovers 50% of stock, cardigans 50% of stock

Lower: 300 units × 75% = _225_ units pullovers
300 units × 25% = _75_ units cardigans

Middle: 536 units × 60% = _322_ units pullovers
536 units × 40% = _214_ units cardigans

High: 250 units × 50% = _125_ units pullovers
250 units × 50% = _125_ units cardigans

Once the buyer has determined the overall units, we need to further break down the styles and tentative sizes.* For example, after further analyzing the fashion trends, the buyer determines that pullovers should be crewneck and V-neck and cardigans should be divided between sleeveless vests and traditional V-neck cardigans.

---

*At this point, the stock should be broken down by color and size. All major manufacturers and designers will advise merchandisers and business owners that some sizes sell better in some areas of the country than will other sizes. This decision is, of course, based on styling and fit; however, it is important to remember that all assortments do not have the same number of pieces.

For pullovers: Trends indicated that crewnecks and V-necks are equally popular, so the split will be 50% in each category. We enter the following calculations in the chart on page 199. When we need to round a number such as 112.5, we choose whether to round up or down. That is, we could just as well have written 113 V-necks and 112 crewnecks in the lower-price group.

For cardigans: Based on information from the corporate buying offices and manufacturers, the designer determines that, in both the lower and middle price lines, 25% should be traditional V-necks and 75% should be sleeveless vests. In the upper price line, 35% should be traditional and 65% should be sleeveless.

Lower price: 225 units × 50% = ___112___   V-neck
                 225 units × 50% = ___113___   crewneck
                 75 units × 25% = ___19___   traditional
                 75 units × 75% = ___56___   sleeveless

Middle price: 322 units × 50% = ___161___   V-neck
                 322 units × 50% = ___161___   crewneck
                 214 units × 25% = ___54___   traditional
                 214 units × 75% = ___160___   sleeveless

Upper price: 125 units × 50% = ___62___   V-neck
                 125 units × 50% = ___63___   crewneck
                 125 units × 35% = ___44___   traditional
                 125 units × 65% = ___81___   sleeveless

Classification Planning Chart

Classification ___Sweaters, class 8___ (Name and identifying number)

Season or month ___November___ Total BOM stock ___$250,000___

Classification % to total stock ___12%___ Total planned dollars ___$30,000___

	Price Line	% Value	Total Retail $	Total Units
Lower	$20.00	20%	$ 6,000	300 pieces
Middle	$28.00	50%	$15,000	536 pieces
Upper	$36.00	30%	$ 9,000	250 pieces

Inventory Breakdown

Lower		Middle		Upper	
Units	Pieces/units	Units	Pieces/units	Units	Pieces/units

Lower

Style	Style
Pullover	Cardigan
75%	25%
225 pieces/units	75 pieces/units

Middle

Style	Style
Pullover	Cardigan
60%	40%
322 pieces/units	214 pieces/units

Upper

Style	Style
Pullover	Cardigan
50%	50%
125 pieces/units	125 pieces/units

	Subclassification				Subclassification				Subclassification			
	V	C	T	S	V	C	T	S	V	C	T	S
	50%	50%	25%	75%	50%	50%	25%	75%	50%	50%	35%	65%
	112 pcs	113 pcs	19 pcs	56 pcs	161 pcs	161 pcs	54 pcs	160 pcs	62 pcs	63 pcs	44 pcs	81 pcs
Small	___	___	___	___	___	___	___	___	___	___	___	___
Med.	___	___	___	___	___	___	___	___	___	___	___	___
Large	___	___	___	___	___	___	___	___	___	___	___	___
X-Large	___	___	___	___	___	___	___	___	___	___	___	___

Now you can break the units into sizes and sizes into colors. $30,000 no longer seems to buy a large amount of stock.

■ Problems

1. Using the following classification plan, prepare an assortment plan for men's undershirts which are classification 26 of the men's furnishings department. Use the following information and research to assist you in the development of your plan.

   Men's furnishings BOM stock is $45,000 for the spring season.
   Past sales indicate undershirts account for 35% of the sales, and the stock has been on hand to support those sales.

   Key price points are:

   - Lower: $5.00 per package, contributing 20% of the stock
   - Middle: $8.00, contributing 65% of the stock
   - Upper: $12.00, contributing 15% of the stock

   Further research shows:

   - Small sizes contribute 15% of the stock.
   - Medium sizes contribute 30% of the stock.
   - Large sizes contribute 40% of the stock.
   - X-large sizes contribute 15% of the stock.

   Styles are:

   - V-neck and round neck, both of which sell 50–50 in each price range

   Colors are:

   - 100% white in the lower and medium price points
   - 80% white and 20% fashion colors in the upper range

Classification Planning Chart

Classification _____ (Name and identifying number)

Season or month _____ Total BOM stock _____

Classification % to total stock _____ Total planned dollars _____

	Price Line	% Value	Total Retail $	Total Units
Lower	_____	_____	_____	_____
Middle	_____	_____	_____	_____
Upper	_____	_____	_____	_____

Inventory Breakdown

Lower pieces/units _____ Middle pieces/units _____ Upper _____ pieces/units

Lower		**Middle**		**Upper**	
Round neck	V-neck	Round neck	V-neck	Round neck	V-neck
_____	_____	_____	_____	_____	_____

Lower (white only)

Small __ % __ pieces   round neck (50%) _____ pcs. v-neck (50%) _____ pcs.
Medium __ % __ pieces   round neck (50%) _____ pcs. v-neck (50%) _____ pcs.
Large __ % __ pieces   round neck (50%) _____ pcs. v-neck (50%) _____ pcs.
Xlarge __ % __ pieces   round neck (50%) _____ pcs. v-neck (50%) _____ pcs.

Middle (white only)

Small __ % __ pieces   round neck (50%) _____ pcs. v-neck (50%) _____ pcs.
Medium __ % __ pieces   round neck (50%) _____ pcs. v-neck (50%) _____ pcs.
Large __ % __ pieces   round neck (50%) _____ pcs. v-neck (50%) _____ pcs.
Xlarge __ % __ pieces   round neck (50%) _____ pcs. v-neck (50%) _____ pcs.

Upper (white) 80% __ pieces

Small __ % __ pieces   round neck (50%) _____ pcs. v-neck (50%) _____ pcs.
Medium __ % __ pieces   round neck (50%) _____ pcs. v-neck (50%) _____ pcs.
Large __ % __ pieces   round neck (50%) _____ pcs. v-neck (50%) _____ pcs.
Xlarge __ % __ pieces   round neck (50%) _____ pcs. v-neck (50%) _____ pcs.

Upper (colors) 20% __ pieces

Small __ % __ pieces   round neck (50%) _____ pcs. v-neck (50%) _____ pcs.
Medium __ % __ pieces   round neck (50%) _____ pcs. v-neck (50%) _____ pcs.
Large __ % __ pieces   round neck (50%) _____ pcs. v-neck (50%) _____ pcs.
Xlarge __ % __ pieces   round neck (50%) _____ pcs. v-neck (50%) _____ pcs.

2. Using the following classification plan, prepare an assortment plan for wallets, which are classification #12 of the women's handbag department. Use the following information and research to assist you in the development of your plan.

BOM stock of women's handbags for the opening of the fall season is $250,000.

Past sales indicate that wallets account for 25% of the sales, and the stock has been on hand to support those sales.

Key price points are:

- Lower: $15.00 per package, contributing 25% of the stock
- Middle: $20.00, contributing 40% of the stock
- Upper: $35.00, contributing 35% of the stock

Further research shows that 85% of the stock on hand is leather, and the other stock is miscellaneous fabrics.

Since wallets do not have sizes such as small, medium, and large, what are some of the key characteristics you might want to consider when developing a model stock for this classification?

Classification Planning Chart

Classification _____ (Name and identifying number)

Season or month _____ Total BOM stock _____

Classification % to total stock _____ Total planned dollars _____

	Price Line	% Value	Total Retail $	Total Units
Lower	_____	_____	_____	_____
Middle	_____	_____	_____	_____
Upper	_____	_____	_____	_____

Inventory Breakdown

Lower
Leather _____ % _____ pcs.   Fabric _____ % _____ pcs.

Middle
Leather _____ % _____ pcs.   Fabric _____ % _____ pcs.

Upper
Leather _____ % _____ pcs.   Fabric _____ % _____ pcs.

3. Using the following classification plan, prepare an assortment plan for men's jogging shorts in a sporting goods store, which are classification 18 of the men's exercise department. Use the following information and research to assist you in the development of your plan.

The BOM stock for the exercise department on May 1 is $385,000.

Past sales indicate jogging shorts account for 30% of the sales, and the stock has been on hand to support those sales.

Fabrication and colors are key factors for the consumer, and the buyer must consider the care and durability of the product also, but this breakdown will not be as detailed because availability from the manufacturer will be considered when purchasing.

Key price points are:

- Lower: $16.00 per package, contributing 30% of the stock
- Middle: $24.00, contributing 60% of the stock
- Upper: $32.00, contributing 10% of the stock

Further research shows:

- Small sizes contribute 15% of the stock.
- Medium sizes contribute 20% of the stock.
- Large sizes contribute 40% of the stock.
- X-large sizes contribute 25% of the stock.

Classification Planning Chart

Classification _____ (Name and identifying number)

Season or month _____ Total BOM stock _____

Classification % to total stock _____ Total planned dollars _____

	Price Line	% Value	Total Retail $	Total Units
Lower	_____	_____	_____	_____
Middle	_____	_____	_____	_____
Upper	_____	_____	_____	_____

Inventory Breakdown

Lower

_____ total pieces

Middle

_____ total pieces

Upper

_____ total pieces

Small ____ % ____ pieces   Small ____ % ____ pieces   Small ____ % ____ pieces

Medium ____ % ____ pieces   Medium ____ % ____ pieces   Medium ____ % ____ pieces

Large ____ % ____ pieces   Large ____ % ____ pieces   Large ____ % ____ pieces

X-large ____ % ____ pieces   X-large ____ % ____ pieces   X-large ____ % ____ pieces

Stock and Sales Planning by Classification

Sometimes when planning assortments, the buyer will focus on the sales generated instead of the stock. In the following problems, use the same steps you just went through, but this time keep in mind the buyer is looking at sales and how many pieces he or she should buy instead of how the stock should be developed.

4. The woman's lingerie buyer is planning the fall season. Based on historical data, the buyer calculates that 15% of the department sales will be generated from new high-fashion cotton boxers. Boxers are classified as no. 15 on the sales and classifications reports. Department sales are planned at $540,000. The following are the buyer's anticipated sales plans. In this assortment there will be two price lines, one at $4.00 and one at $6.00. The $4.00 price line will generate 60% of the sales, and 40% will come from the $6.00 price line. Use the following size breakdown to determine pieces per size:

Small: 10%
Medium: 30%
Large: 40%
X-large: 20%

Classification Planning Chart

Classification _____ (Name and identifying number)

Season or month _____ Total planned dollars _____

Price Line	% Value	Total Retail $	Total Units
$4.00	_____	_____	_____
$6.00	_____	_____	_____

Inventory Breakdown

$4.00

____ pcs./units

Small ____ % ____ pieces

Medium ____ % ____ pieces

Large ____ % ____ pieces

Xlarge ____ % ____ pieces

$6.00

____ pcs./units

Small ____ % ____ pieces

Medium ____ % ____ pieces

Large ____ % ____ pieces

Xlarge ____ % ____ pieces

5. Based on the following information, develop a classification chart that you feel will be useful for determining a successful assortment. The children's wear buyer plans a 6-month sales goal of $820,000. From this sales volume, 20% will come from boys' size 4–7 pants. Of that volume, 70% will come from casual pants and 30% will come from dressy (uniform) pants.

In the casual subclass, note these price points:

$8.00,   20%

$10.00,   50%

$12.00,   30%

In the dressy subclass, note these price points:

$10.00,   30%

$12.00,   50%

$15.00,   20%

Of the entire inventory 60% is in regular sizes and 40% is in slim sizes. Further size breakdown is as follows:

	Regular	Slim
Size 4	10%	None
Size 5	20%	30%
Size 6	40%	50%
Size 7	30%	20%

Classification Planning Chart

Classification _____ (Name and identifying number)

Season or month _____ Total BOM Stock _____

Classification % to total stock _____ Total planned dollars _____

Casual Pants Classification

Price Line	% Value	Total Retail $	Total Units
$8.00	_____	_____	_____
$10.00	_____	_____	_____
$12.00	_____	_____	_____

$8.00
Regular ___ % ___ pieces
4 ___ % ___ pcs.
5 ___ % ___ pcs.
6 ___ % ___ pcs.
7 ___ % ___ pcs.

$10.00
Regular ___ % ___ pieces
4 ___ % ___ pcs.
5 ___ % ___ pcs.
6 ___ % ___ pcs.
7 ___ % ___ pcs.

$12.00
Regular ___ % ___ pieces
4 ___ % ___ pcs.
5 ___ % ___ pcs.
6 ___ % ___ pcs.
7 ___ % ___ pcs.

$8.00
Slim ___ % ___ pieces
4 ___ % ___ pcs.
5 ___ % ___ pcs.
6 ___ % ___ pcs.
7 ___ % ___ pcs.

$10.00
Slim ___ % ___ pieces
4 ___ % ___ pcs.
5 ___ % ___ pcs.
6 ___ % ___ pcs.
7 ___ % ___ pcs.

$12.00
Slim ___ % ___ pieces
4 ___ % ___ pcs.
5 ___ % ___ pcs.
6 ___ % ___ pcs.
7 ___ % ___ pcs.

Dressy Pants Classification

Price Line	% Value	Total Retail $	Total Units
$10.00	_____	_____	_____
$12.00	_____	_____	_____
$15.00	_____	_____	_____

$10.00
Regular ___ % ___ pieces
4 ___ % ___ pcs.
5 ___ % ___ pcs.
6 ___ % ___ pcs.
7 ___ % ___ pcs.

$12.00
Regular ___ % ___ pieces
4 ___ % ___ pcs.
5 ___ % ___ pcs.
6 ___ % ___ pcs.
7 ___ % ___ pcs.

$15.00
Regular ___ % ___ pieces
4 ___ % ___ pcs.
5 ___ % ___ pcs.
6 ___ % ___ pcs.
7 ___ % ___ pcs.

$10.00
Slim ___ % ___ pieces
4 ___ % ___ pcs.
5 ___ % ___ pcs.
6 ___ % ___ pcs.
7 ___ % ___ pcs.

$12.00
Slim ___ % ___ pieces
4 ___ % ___ pcs.
5 ___ % ___ pcs.
6 ___ % ___ pcs.
7 ___ % ___ pcs.

$15.00
Slim ___ % ___ pieces
4 ___ % ___ pcs.
5 ___ % ___ pcs.
6 ___ % ___ pcs.
7 ___ % ___ pcs.

Profitability

Operating Statements

Purpose of an Operating Statement

Once merchants have set the department plans or store plans into motion for a season, they follow a specific set of activities. Merchandise is

Purchased with the best possible terms

Priced for the greatest profits

Promoted with creativity

At this point it is time to determine if the company actions have been profitable.

The three major components to review are the following:

1. Sales volume
2. Cost of goods sold
3. Operating expenses

These components, which are found in financial statements, are tools of comparison to use in determining how a business is operating. Corporate goals are about earning profits. Simply stated, an operating statement allows a business owner to compare a variety of factors and use these statistics as a guide to improve, build, or restructure whatever may be needed for building a profitable operation.

Industry **Terms** and *jargon*

Operating Statement   A detailed financial statement that reviews income and expenses for a retail operation, sometimes called a *profit and loss statement.* This financial report is detailed, outlining specific components.

Income Statement   A financial review of income and expenses. In retail businesses, it is also called a *skeletal P/L*.

Skeletal Profit and Loss Statement   A financial review of business income and expenses. This differs from an operating statement in that it does not detail specific areas. Some businesses prepare a skeletal P/L monthly and refer to it as an *income statement.*

Quarter   A 13-week period. There are four quarters in a year. At the end of a quarter, a business examines the financial results for that period. (Review Chapter 7.)

Annual   A 1-year period of time from January 1 through December 31. (Review Chapter 7.)

Fiscal Year   An accounting period of 12 months. In the merchandising market, the fiscal year runs from February 1 through January 31. (Review Chapter 7.)

Gross Sales   All sales made in a specific period of time. (Review chapters 2 and 7.)

Customer Returns and Allowances   Credits issued for merchandise brought back after a sale or credit allowed for a defect or price adjustment (allowance). (Review chapters 2 and 7.)

Net Sales   Total final sales for a period of time. Gross sales minus customer returns and allowances equals net sales. (Review Chapters 2 and 7.)

Opening Inventory at Cost   The cost value of the merchandise (inventory) on hand at the beginning of the month (or period of time being reviewed). This can be determined in two different ways: (1) The value determined by a physical count and recorded as such or (2) a review of the BOM dollar retail inventory multiplied by the markup complement.

$$100\% - MU\% = C\%$$
$$R\$ \times C\% = C\$$$

(Review Chapter 4.)

Physical Inventory   The specific amount of merchandise counted at least two times a year. (Review Chapter 7.)

Book Inventory   The financial bookkeeping records of the inventory based on a physical starting count plus purchases minus sales, stock reductions, and employee discounts. (Review Chapter 7.)

Purchases at Cost   Value of the merchandise at cost that has been received during the financial period in review. (Review Chapter 7.)

Return to Vendor (RTV)   Also called *charge-backs,* or *purchase returns,* the dollar value of merchandise sent back to a manufacturer and credit applied to the department inventory on hand. (Review Chapter 3.)

Transportation Charges   Shipping expenses. (Review Chapter 3.)

Total cost of purchases   The value is determined by adding the purchases and the transportation charges.

Closing Inventory at Cost   The value of the merchandise at the end of the period in review. This figure is determined by identifying the retail stock, either with a physical count or by a review of six-month financial records of actual figures and calculating the cost value:

$$100\% - \text{department MU} = \text{cost } \%$$
$$\text{Retail inventory} \times \text{C}\% = \text{cost inventory}$$

For example:
MU = 52 %
100% − 52% = 48%
Retail inventory = $106,000
$106,000 × 48% =
$50,880 at cost.

(Review Chapters 4 and 7.)

Gross Cost of Merchandise Sold (CMS)   The cost value of the merchandise that was purchased by customers. This is determined as follows:

$$\begin{array}{rl} & \text{Opening inventory at cost} \\ + & \text{Purchases at cost} \\ - & \text{Closing inventory at cost} \\ \hline = & \text{Cost of the merchandise sold} \end{array}$$

Cost of Goods Sold   Term used interchangeably with *cost of merchandise sold.*

Cash Discounts   Discounts earned for prompt payment. (Review Chapter 3.)

Net Cost of Merchandise Sold (CGS)   The value of the gross cost of merchandise sold minus the cash discounts earned.

Workroom Expenses   Costs incurred to complete a sale. (Review Chapter 7.)

Total CMS or Total CGS   The final value is determined as follows:

$$\begin{array}{rl} & \text{Gross cost of goods sold} \\ - & \text{Cash discounts earned} \\ + & \text{Workroom expenses} \\ \hline = & \text{Total cost of goods sold} \end{array}$$

Gross Margin   Net sales minus the cost of goods sold. This is sometimes referred to as *gross profit.*

Operating Expenses   Costs incurred in running a business operation. (Review Chapter 4.)

Direct Expenses   Expenses that directly affect the operations of the area being reviewed, such as salespeople's salaries for the department, advertising for the department, or the buyer's salary for the department.

Indirect Expenses   Expenses that exist even if a department is not part of the overall operation. For example, a store pays rent. The store controller identifies how much physical floor space is used by the department. That percentage is then applied to the rent, and that proportion of the rent is charged to the department. Also, a percentage of the salaries of the officers of a corporation is distributed in relationship to the volume that a department earns. In simple terms, every department contributes some of its earnings toward the salaries of the chief operating officers of a company.

Contribution Margin   The amount remaining after direct expenses are deducted from the gross margin. Although this term is not used frequently, businesses preparing detailed operating statements will include it in the final reports. This figure represents how much money is still left to contribute to overall store expenses. From this value the indirect expenses are deducted to then yield the operating profit or loss, as you see below:

> Net sales
> −   Cost of merchandise sold
>
> =   Gross margin
> −   Direct expenses
>
> =   Contribution margin
> −   Indirect expenses
>
> =   Net operating profit (or loss)

Net Operating Profit (or Loss)   The difference between gross margin and total operating expenses. Additional expenses, such as: interest on loans, income taxes, or unpaid receivables (also called bad debts), need to be deducted.

Net Profit   The net operating profit with all additional expenses (i.e., interest, taxes) deducted.

Operating Loss   Also called *net loss,* the amount of loss when operating expenses exceed gross margin.

YTD   Year to date. Financial records are often done in two ways. A monthly recap is prepared to allow a business to focus in on the specific activities for a period. That information is combined with previous data for a YTD overview. For example, if a statement is dated Month of June 30, 199—, it represents the financial activities for that specific month. If the statement says YTD June 30, 199—, it means that the figures represent the financial transactions from the beginning of the fiscal year through June 30, 199—.

Format of an Operating Statement

The following shows the format of a detailed operating (P/L) statement.

	Dollars	Percents
Gross sales		
− Customer returns and allowance		
= Net sales	_____	100 %
Opening inventory at cost		
+ Purchases at cost		
+ Transportation charges		
= Total goods handled at cost		
− Closing inventory		
= Gross cost of goods sold		
+ Workroom charges (if applicable)		
− Cash discounts earned		
= Net cost of goods sold	− _____	
= Gross margin	_____	
Direct Expenses		
Buying salary		
+ Selling salaries		
+ Dept. advertising		
+ Dept. receiving expenses		
= Total direct expenses	− _____	%
= Contribution margin	_____	%
Indirect Expenses		
Allocated rent		
+ Allocated exec. salaries		
+ Allocated utilities		
+ Allocated maintenance		
= Total indirect expenses	− _____	%
= Net profit (or loss)	_____	%

If a business wants a quick overview, it will prepare a skeletal profit and loss statement. The skeletal profit and loss statement is also called an income statement. These statements can be set up in a basic T-chart grid to analyze the information effectively in both dollars and percents.

	Dollars	**Percents**
Net sales		100%
− CGS		
= Gross margin		
− Operating expenses		
= Net profit (loss)		

Let's take a look at an operating statement that has been completed.

		Dollars	Percents
Gross sales		$685,000	
− Customer returns and allowance		$ 12,850	
	= Net sales	$672,150	100%
Opening inventory at cost		$84,000	
+ Purchases at cost		$195,000	
+ Transportation charges		$14,200	
= Total goods handled at cost		$293,200	
− Closing inventory at cost		−$107,000	
= Gross cost of goods sold		$186,200	
+ Workroom charges (if applicable)		none	
− Cash discounts earned		−$15,600	
Net cost of goods sold		$170,600 − $170,600	25.38%
	= Gross margin	501,550 (74.62%)	
Direct Expenses			
Buying salary		$62,000	
+ Selling salaries		$93,750	
+ Dept. advertising		$24,000	
+ Dept. receiving expenses		$16,200	
= Total direct expenses		$195,950 −$195,950	29.15%
	= Contribution margin	305,600	45.47%
Indirect Expenses			
Allocated rent		$72,000	
+ Allocated exec. salaries		$116,000	
+ Allocated utilities		$35,000	
+ Allocated maintenance		$27,000	
= Total indirect expenses		$250,000 −$250,000	37.20%
	= Net profit (or loss)	$55,600	8.27%

Helpful Hint 2
$501,550 \div 672,150 = 74.62\%$

Don't forget your Helpful Hints and remember to check your work.

> *Helpful Hint 1*
> **Total amount $\times$ percent = specific amount**
>
> *Helpful Hint 2*
> **Specific amount $\div$ total amount = percent**
>
> *Helpful Hint 3*
> **Specific amount $\div$ percent = total amount**
>
> *Helpful Hint 4*
> **Difference $\div$ original amount = percent of increase (or decrease)**

Now, let's take a look at a completed skeletal profit and loss statement.

	Dollars	**Percents**
Net sales	852,000	100%
− CGS	625,000	73.36%
= Gross margin	227,000	26.64%
− Operating expenses	243,000	28.52%
= Net profit (loss)	(16,000)	(1.88%) **Loss**

Step 1 $ 852,000 $625,000 ÷ $852,000 = 73.36%

          − 625,000 $227,000 ÷ $852,000 = 26.64%

          $ 227,000   Check your work: 100% − 73.36% = 26.64%

Step 2 $ 227,000 $243,000 ÷ $852,000 = 28.62%

          − 243,000 $16,000 ÷ $852,000 = 1.88%

          ($16,000)   Remember: Check your work.

Determining Net Cost

It is important to be able to determine the net cost of the merchandise before you start completing a detailed profit and loss statement. Here is an example.

EXAMPLE

Given the following information:

Billed cost of goods	$3,000
Transportation costs	$ 204
Cash discounts earned	$ 90

Determine the net cost of merchandise.

$$\begin{array}{r} \$3,000 \\ +\quad 204 \\ \hline \$3,204 \\ -\quad 90 \end{array}$$

3000 $\boxed{+}$ 204 $\boxed{-}$ 90 $\boxed{=}$ 3114

Net cost of merchandise = $3,114.   ■

■ Problems

When analyzing profit and loss statements, always try to do the following:

1. Set up a basic T-chart using net sales as 100%.
2. If gross sales are indicated, deduct the customer returns and allowances and determine net sales *before* beginning to prepare the basic skeletal profit and loss (income) statement.
3. Follow the *Helpful Hints* outlined in Chapter 2.

Skeletal P/L Statements

1. Set up a skeletal profit and loss statement using the following data:

Net sales = $35,000

Cost of merchandise sold = $18,000

Expenses = $7,200

	Dollars	Percents
Net sales		100%
− CMS		
= Gross margin		
− Operating expenses		
= Profit (or loss)		

2. Set up a skeletal profit and loss statement using the following data:

Net sales = $285,000

Cost of merchandise sold = $209,000

Expenses = 39.5%

	Dollars	Percents
Net sales		100%
− CMS		
= Gross margin		
− Operating expenses		
= Profit (or loss)		

3. Set up a skeletal profit and loss statement using the following data:

$$\text{Gross margin} = \$147{,}000$$
$$\text{Expenses} = \$149{,}000$$
$$\text{Loss} = (0.5\%)$$

	Dollars	Percents
Net sales		100%
− CMS		
= Gross margin		
− Operating expenses		
= Profit (or loss)		

4. Set up a skeletal profit and loss statement using the following data:

$$\text{CMS} = \$260{,}000$$
$$\text{Expenses} = \$240{,}000$$
$$\text{Profit} = \$20{,}000$$

	Dollars	Percents
Net sales		100%
− CMS		
= Gross margin		
− Operating expenses		
= Profit (or loss)		

5. Set up a skeletal profit and loss statement using the following data:

Gross sales = $162,000 Customer R/A = $21,000
Expenses   = $27,500   Gross margin = $54,000

	Dollars	Percents
Net sales		100%
− CMS		
= Gross margin		
− Operating expenses		
= Profit (or loss)		

6. Set up a skeletal profit and loss statement using the following data:

Gross sales = $115,000 Customer R/A = $1,450
Expenses   = $37,500   Loss   = $2,650

	Dollars	Percents
Net sales		100%
− CMS		
= Gross margin		
− Operating expenses		
= Profit (or loss)		

Net Cost

7. Determine the net cost of the merchandise purchased for the department.

Billed cost of goods	$7,000
Transportation costs	$ 100
Alteration charges	$ 90
Cash discounts earned	$ 80

The net cost of merchandise is _____.

8. Using the net cost of merchandise in Problem 7, determine the gross margin if sales are $11,560.

9. Peter and Ryan's Car Stop recorded the following figures for the month of April:

Net sales	$14,280
Billed cost of goods	$ 7,620
Transportation costs	$ 120
Alteration charges	$ 105
Cash discounts earned	$ 405

The gross margin for April is _____.

10. Jammer's reported the following figures for the month of February for the videotape department.

Net sales	$56,000
Billed cost of goods	$40,000
Transportation costs	$ 1,000
Cash discounts earned	$ 500

The gross margin for February is _____.

11. The sportswear department at Helen's Haberdashery reported the following figures for the month of September:

Net sales	$140,000
Billed cost of goods	$100,000
Transportation costs	$ 1,502
Cash discounts earned	8% of the billed cost of goods

The gross margin for September in the sportswear department is _____.

Detailed P/L Statements

12. a. Use a detailed operating statement to determine the dollar values and percentages for this business. Decide if the business is operating at a profit or loss and by how much in both dollars and percentages.

b. Write a paragraph outlining areas that you would like to have additional information about and review for more detail. Explain your choices.

Gross sales	$270,000
Customer returns and allowances	$ 25,000
Opening inventory	$ 99,000
Purchases at cost	$155,000
Transportation charges	$ 2,500
Ending inventory at cost	$122,000
Workroom expenses	$ 4,000
Cash discounts earned	$ 12,400
Direct Expenses	
Buying salaries	$ 8,000
Selling salaries	$ 13,000
Dept. advertising	$ 18,000
Warehouse expenses	$ 5,000
Indirect Expenses	
Executive salaries	$ 19,000
Rent	$ 18,500
Utilities	$ 12,000
Maintenance	$ 7,800

		Dollars	Percents
Gross sales			
− Customer returns and allowance			
	= Net sales	_____	100%
Opening inventory at cost			
+ Purchases at cost			
+ Transportation charges			
= Total goods handled at cost			
− Closing inventory at cost			
= Gross cost of goods sold			
+ Workroom charges			
− Cash discounts earned			
Net cost of goods sold		−_____	%
	=Gross margin	_____	%
Direct Expenses			
Buying salary			
+ Selling salaries			
+ Dept. advertising			
+ Dept. receiving expenses			
= Total direct expenses		−_____	%
	=Contribution margin		%
Indirect Expenses			
Allocated rent			
+ Allocated exec. salaries			
+ Allocated utilities			
+ Allocated maintenance			
= Total indirect expenses		−_____	%
	=Net profit (or loss)	_____	%

13. a. Use a detailed operating statement to determine the dollar values and percentages for this business. Determine if the business is operating at a profit or loss and by how much in both dollars and percentages.

b. Write a paragraph outlining areas that you would like to get additional information about and review for more detail. Explain your choices.

Gross sales	$909,290
Customer returns and allowances	$ 51,466
Cost of merchandise sold	$471,342

Direct Expenses

Buying salaries	$ 9,300
Selling salaries	$ 55,170
Dept. advertising	$ 24,876
Warehouse expenses	$ 7,145

Indirect Expenses

Executive salaries	$ 39,060
Rent	$ 25,360
Utilities	$ 25,430
Maintenance	$ 16,880

Note: The cost of merchandise sold in this problem has been determined for you.

	Dollars	Percents
Gross sales		
− Customer returns and allowance		
= Net sales	_____	100%
Opening inventory at cost		
+ Purchases at cost		
+ Transportation charges		
= Total goods handled at cost		
− Closing inventory at cost		
= Gross cost of goods sold		
+ Workroom charges		
− Cash discounts earned		
Net cost of goods sold	− _____	%
= Gross margin	_____	%
Direct Expenses		
Buying salary		
+ Selling salaries		
+ Dept. advertising		
+ Dept. receiving expenses		
= Total direct expenses	− _____	%
= Contribution margin		%
Indirect Expenses		
Allocated rent		
+ Allocated exec. salaries		
+ Allocated utilities		
+ Allocated maintenance		
= Total indirect expenses	− _____	%
= Net profit (or loss)	_____	%

14. Use a detailed operating statement to determine the dollar values and percentages for this business. Decide if the business is operating at a profit or loss and by how much in both dollars and percentages.

Gross sales	$130,005
Customer returns and allowances	$ 10,850
Opening inventory at cost	$ 51,600
Purchases at cost	$ 80,400
Closing inventory at cost	$ 59,760
Transportation charges	$ 2,000
Workroom expenses	$ 1,200
Buying salary	$ 4,500
Selling salaries	$ 11,480
Rent	$ 9,450
Executive salaries	$ 12,250

	Dollars	Percents
Gross sales		
− Customer returns and allowance		
= Net sales	_____	100%
Opening inventory at cost		
+ Purchases at cost		
+ Transportation charges		
= Total goods handled at cost		
− Closing inventory at cost		
= Gross cost of goods sold		
+ Workroom charges		
− Cash discounts earned		
Net cost of goods sold	−_____	
=Gross margin	_____	
Direct Expenses		
Buying salary		
+ Selling salaries		
+ Dept. advertising		
+ Dept. receiving expenses		
= Total direct expenses	−_____	%
=Contribution margin		%
Indirect Expenses		
Allocated rent		
+ Allocated exec. salaries		
+ Allocated utilities		
+ Allocated maintenance		
= Total indirect expenses	−_____	%
=Net profit (or loss)	_____	%

15. Use a detailed operating statement to determine the dollar values and percentages for this business. Decide if this business is operating at a profit or loss and by how much in both dollars and percentages.

Gross sales	$102,500
Customer returns and allowances	$ 15,000
Opening inventory at cost	$ 39,000
Purchases at cost	$ 42,400
Closing inventory at cost	$ 36,750
Transportation charges	$ 1,250
Cash discounts earned	$ 1,200
Buying salary	$ 4,500
Selling salaries	$ 5,450
Rent	$ 5,500
Department advertising	$ 6,500
Executive salaries	$ 7,250
Misc. (maintenance, supplies, etc.)	$ 10,000

	Dollars	Percents
Gross sales		
− Customer returns and allowance		
=Net sales	_____	100%
Opening inventory at cost		
+ Purchases at cost		
+ Transportation charges		
= Total goods handled at cost		
− Closing inventory at cost		
= Gross cost of goods sold		
+ Workroom charges		
− Cash discounts earned		
Net cost of goods sold	−_____	
=Gross margin	_____	
Direct Expenses		
Buying salary		
+ Selling salaries		
+ Dept. advertising		
+ Dept. receiving expenses		
= Total direct expenses	−_____	%
=Contribution margin		%
Indirect Expenses		
Allocated rent		
+ Allocated exec. salaries		
+ Allocated utilities		
+ Allocated maintenance		
= Total indirect expenses	−_____	%
=Net profit (or loss)	_____	%

fourteen

Sales Per Square Foot

As business managers evaluate a department's or store's performance, one of the key determinations of profitability is **sales per square foot.** Calculating sales per square foot gives the value of the sales generated in relationship to the amount of floor space. Many chain store operations develop a **planogram** in an effort to

- Promote merchandise.
- Maintain unified "looks" in all stores.
- Keep a traffic flow that encourages consumer traffic.
- Maintain an efficient stock area.
- Make sure their operation is generating enough volume.
- Judge sales per square foot in relationship to leasing costs per square foot.*

These planograms are also called **prototypes.** Visual coordinators prepare a store layout with in-store displays and window layouts in a mock store setting. These layouts are photographed, and a *scaled* floor plan is prepared and sent to all stores, along with specific dates of layout changes. These planograms, or prototypes, have a twofold purpose:

1. To present an attractive merchandise mix promoting significant vendors and/or fashion trends
2. To allow a business owner to compare and evaluate the use of space in order to utilize it more effectively

---

*Leasing agreements are done by the square foot. If a business owner plans to open a new store, part of analyzing the site location will involve reviewing how much a leasing firm charges per square foot to rent a store space.

Industry **Terms** and *jargon*

Planogram   A drawn presentation of store layout identifying vendor space, merchandise trends, signing, and the store's visual displays. Also called *prototypes*.

Sales Per Square Foot   A figure that identifies how effective an area of floor space is in relationship to the total store area.

$$\text{sales} \div \text{floor space} = \text{sales per square foot}$$

Square Feet   A unit used to measure area. In a store, area is determined by multiplying length by width.

Quarter-inch Scale   A scale in which every $\frac{1}{4}$-inch square on a planogram represents 1 foot of the store plan. An accurate work drawing helps eliminate many on-the-job errors.

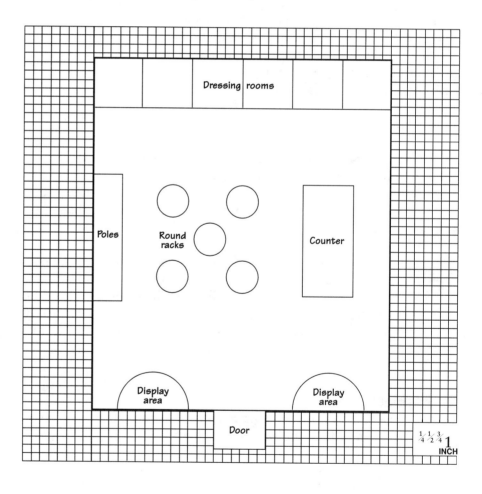

Sales per square foot is used as an indicator to determine how strong a business operation is. If a department is generating high sales per square foot, it is usually one sign of a profitable operation.

The following example shows these ideas.

EXAMPLE

Consider the known information for a small store.

Overall size: 30 ft by 50 ft
Stock area in back: 30 ft by 4 ft
Dressing room and checkout: 6 ft by 20 ft
Display area in front: 3 ft by 50 ft
November sales: $32,000

Find the actual square footage of the selling space and determine sales per square foot.

Overall size:	30 × 50 = 1,500	1,500
Stock area in back:	– 30 × 4 = 120	– 120
Dressing room and checkout	– 6 × 20 = 120	– 120
Display area	– 3 × 50 = 150	– 150
	1,110 **Selling area**	1,110

32000 ⊟ 1110 ⊟ 28.83 $32,000 ÷ 1,110 = $28.83

Sales per square foot equal $28.83.

Talk Out

If you are paying $18.00 a square foot to lease space, what do you think?   ■

■ Problems

1. Determine the annual sales per square foot.

Department store
Accessories area: 12 ft × 8 ft
Annual sales: $82,500
Displays: incorporated in the fixtures
Dressing area: none

2. Sales for the dress department in a downtown store are $250,000. The square footage for the department is 1,512. What are the sales per square foot to the nearest cent?

3. The sales for the department are $178,500. The department area is 22 ft by 44 ft. What are the sales per square foot to the nearest cent?

4. Sales volume for a children's area covering infants through boys 8–20 is $3,500,000 annually. The area covers 70,000 ft^2. Find the sales per square foot.

5. Determine annual sales per square foot.

   Specialty store: sweater division
   Sweater area: 35 ft × 15 ft
   Annual sales for store: $1,568,000
   Sweater classification: 11% of sales

6. Find the annual sales per square foot.

   Golf department of a sporting goods store
   Department area: 35 ft × 14 ft
   Annual sales: $718,000

7. For a boys 8–20 department:

   Overall size: 65 ft by 58 ft
   Stock area in back: 10 ft by 4 ft
   Dressing room and checkout: 3 ft by 29 ft
   Display area in front: 3 ft by 20 ft
   Annual sales: $1,542,000

   Find the annual sales per square foot. _____
   Find the average monthly sales per square foot. _____

8. For a women's robe department:

   Total size of specialty store: 60 ft × 80 ft

   Overall department size: 15 ft by 10 ft

   Stock area in back for women's robes: 4 ft by 4 ft

   Dressing rooms: four, each 3 × 3 (One of the four is charged to the robe department.)

   Display area in front for robes: 3 ft by 4 ft

   Annual robe sales: $965,000

   Find the annual sales per square foot. _____

   Find the average monthly sales per square foot. _____

9.   a. The men's shoe division in a discount store covers an area of 20 feet by 18 feet. The overall department size is 60 feet by 65 feet. Stock area for the men's division is 10 feet by 2 feet. What is the percent of space for the men's division compared to the entire shoe department?

   b. If the spring season sales volume for the men's division is $152,000, what are the average monthly sales per square foot? _____

   c. If the total shoe department generates a volume of $912,000 for the season, discuss the overall performance of the men's department based on the space allocated and the volume generated.

10.   a. The water sports division in the local camping store is allocated a large area during the spring and summer months, covering 25% of the store space, including the stock area. The store is 300 ft by 200 ft. If the spring season sales volume for this division is 1,875,500, what is the average monthly sales per square foot? _____

   b. During the fall season the space is reduced to 7,500 square feet because the demand drops off and sales are $900,000. Using this information, determine the sales per square foot per month for the fall.

   c. Compare the figures for spring and fall sales per square foot. Based on the facts, what do you think about the position management takes to reduce the square footage in the fall?

Corporate Buying and Planning

CHAPTER *fifteen*

Cost Sheets and Pricing

Fashion Trends

During the last decade the retail fashion industry has made dramatic changes in how merchandise is made, where it is made, and where it is sold. No longer is the traditional department store, specialty store, or mass merchant the store of choice. In the 1990s, the tremendous success of discount giants such as Wal-Mart, Kmart, and Target, along with the enormous outlet shopping malls, has forced retailers to travel the world and find products that are new, exciting, and priced right. This means designers no longer hold the only key to the industry; the doors have now opened to allow buyers, corporate buying offices, and manufacturers to develop products unique to their own special niche market.

Designing in the 1990s has taken on a different look. Products are being offered to the public through

Designers

Product development (sourcing) of national name brands

Private labeling

Designers such as Donna Karan, Tommy Hilfiger, and Calvin Klein are finding themselves in competition with the store buyers and merchants, because everyone is aggressively working in the field to develop the best-priced products for their customer.

Buyers not only need to know how to negotiate the terms and conditions of a product being offered at the wholesale or cost price, they are now wearing a second hat, acting as a designer developing products and lines of merchandise for their own stores. Merchandisers now must be able to determine the best price of materials and labor as they design products they feel can be sold successfully in their stores. Suddenly they are no longer just merchants but are

Merchandising designers

Sourcing buyers or

Product-development specialists

Let's go back for a moment. Until the mid-1980s designers and major manufacturers really did control the fashion industry. However, when the United States hit some severe economic times, retailers needed to find products for the consumer that were similar in styling and quality but had lower prices. It became apparent to the large department stores that companies such as Gap and Limited, who made and sold their own products, were experiencing greater profits than companies relying on designer labels. Gap and Limited are examples of stores who employ their own design team, produce their own products, and sell them in a store with the same name as the product. The **private label** in the 1980s made a mark of its own with great success.

As we approach the 21st century, we find that the large specialty stores have consulting groups and design teams working to find the best materials in the world, along with the least expensive labor costs. In addition, designers are selling their names through licensing rights to multinational companies, who are able to manufacture and produce bridge lines of a product far more inexpensively than the designer. The result is a new breed of retailer throughout the United States. Known as **merchandising designers, sourcing buyers,** or **product-development specialists,** these positions are growing in central buying offices, international buying offices, and product-development firms.

Merchants are responsible for creating products for their customers that will be aesthetically appealing, will fit well, will be priced right, will be easy to care for and durable, and will be in the store at the right time. In addition the merchants must wisely market the products in the store to ensure strong name recognition, thereby creating consumer demand.

Macy's and Federated have done an excellent job with their I.N.C., Jennifer Moore, and Charter Club lines. Wal-Mart found an audience with the Kathy Lee line, JCPenney found a new consumer market with the Original Arizona Jean Co., and Saks has a great customer following with Saks Real Clothes. You can find private labels in all the leading department stores, high end to low: Dillards, Marshall Field, Nordstrom, Kmart, Wal-Mart, and Target as well as specialty department stores, such as Niemann Marcus and Lord and Taylor.

Each store has buyers developing products and cutting out the designer and the designer sales department. This means that the products, although almost identical in styling, fabric, and trends, are produced at lower costs. The middle person, the salesperson, and promotion costs have all been eliminated, and retail stores have found that this is a great way to make stronger profits for their firms.

As you take on more responsibilities in your career you may find a job in a buying office, such as Federated or Saks, as a part of their product-development design team, or your career may lead to the design and buying offices for the leading private-label names such as Gap and Limited. Job opportunities are also found in major buying offices, such as Associated Merchandising Corp. (AMC), Frederick Atkins, or Henry Doneger and Associates. These firms are multinational firms with main offices in New York and branches throughout the world in places such as Thailand, Tokyo, Hamburg, and Istanbul, major manufacturing locations. Your job will be one that involves a strong knowledge of consumer and economic trends, raw products, and major manufacturing sources in the global marketplace.

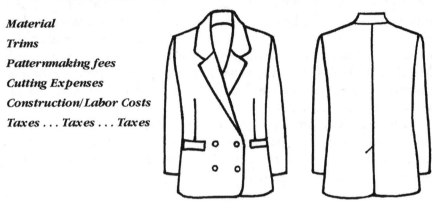

The following charts illustrate the process involved in manufacturing a designer line as opposed to product development.

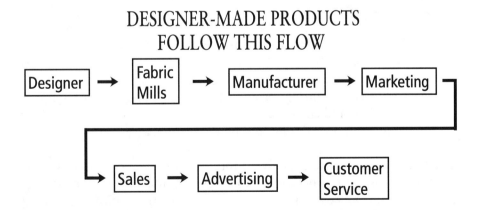

Everyone in this process has to get paid, so the product must be sold for enough to cover the costs to make it and make a profit.

NATIONAL BRAND OR PRIVATE LABEL PRODUCTS FOLLOW THIS FLOW

In this flowchart, the marketing, selling, advertising, and customer service departments are eliminated, and eliminating those expenses keeps the prices down.

In a design firm the designer is right in the middle of the operation, buying the materials to make the product. Once the products are made, buyers can choose what they like, but if a buyer wants to make a product for your store, the trend forecasters, textile expert, and colorist must supply the designer with what the buyer wants to have made. Merchandising designers wear two hats. They have to put the product together, and then they have to make sure that it will be profitable and successfully sell in their stores.

The only way for a designer to do this is to understand how to calculate a cost sheet, which tells exactly how much it will cost to make an item. From that information, a buyer can determine a strong markup. (Review Chapter 4.)

Cost Sheets

When merchandisers and designers create a product or garment for production, even as the first sample is being reviewed, it is essential to prepare a cost sheet that will provide the retail firm or the designer with the **basic elements of cost:**

1. The amount of fabrics and materials required and the cost of these materials
2. The specific trims, findings (i.e., buttons, zippers, elastic, etc.)
3. Labor costs: cutting, production pattern making, grading, marking, and construction
4. Shipping costs, duties, and taxes, if applicable

The total value of these expenses is the *cost of goods to be sold* (CGS).

It is important to know that designers, manufacturers, or merchandisers will always cost out the garment and then establish what they feel is the wholesale, or cost, price. That cost price must be strong enough to support any discounts that the manufacturer or designer will negotiate with the retailer and yet still be competitive.

The following are the elements of a cost sheet form. It is important to note that a cost sheet determines the price of only *one* garment, so let us review how the components work together.

A   The date signifies that the prices placed as of this date are firm. The negotiation and terms of sale have been determined with the textile mills and the findings companies. Generally these prices are good for about 30 days.

B   Style number is the numerical identifier of this garment.

C   Sketch and material swatch of the garment.

D   Description of the garment.

E   Where the fabric is being purchased, important for shipping, timing, duties, import/export quotas, and fees.

F   Where the product is being produced (contractor/country), important because of duties, import/export fees, and quotas.

G   Size range: small, medium, large, for example; size scale: the proportion in which each size is to be cut.

H   Important to know how wide the fabric is so the marker makers can calculate minimum yardage use.

I   Fabric needed for the most effective placement of pattern pieces for a full size run. This is a **critical** number. When all the pattern pieces have been laid out for a complete size run, this is the amount of fabric that is required. The less fabric used, the lower the cost.

J   Actual yardage used. Companies often review cost sheets and spec sheets to determine the actual yardage used to see if their calculations were accurate or if something changed, such as fabric width, that altered the amount of yardage used.

K   Specific yardage for one garment. This is determined by dividing the total marker yardage by the number of sizes being made. For example, "S, M, L, XL; 10.5 yards for our size scale, which is 6 pieces. We're doubling up on mediums and larges (1 small, 2 mediums, 2 larges, 1 extra large)." Although a small size will certainly take less fabric than an extra large, 10.5 yards ÷ 6 = 1.75 yards, so 1.75 is the average yardage identified on the cost sheet.

L   The price negotiated after all the trade discounts are discussed and shipping terms and payment arrangements are made.

M   The total cost of that material (yardage × price per yard).

N   Total cost of all materials for one garment.

O   Number of trim pieces used in one garment.

P   Prices for the trims.

Q   Total cost of the trims (number of pieces × cost).

COST SHEET

Date ___A___

Style number ___B___

Sketch and materials swatch

C

Description ___D___

Fabric mill ___E___

Manufactured ___F___

Size range ___G___

Size scale ___G___

Fabric width ___H___

Estimated marker yardage ___I___

Actual marker yardage ___J___

Materials	Yardage	Price per Yard	Total Amount
Fabric	K	L	M
Fabric			
Lining			
		Total cost of materials	N

Trims	Amounts	Cost Each	Total Amount
Buttons	O	P	Q
Zippers			
Belts			
Other			
		Total cost of trims/findings	R

Labor

Marking costs ___S___   Grading costs ___T___   Cutting costs ___U___

Construction ___V___

Total labor costs ___W___

Shipping and Duty Expenses

Import duties ___X___   Shipping expenses ___Y___

Total shipping and duty costs ___Z___

Total cost of goods to be sold ___AA___

Suggested wholesale price ___BB___

Suggested retail price ___CC___

R   Total cost of all trims and findings for one garment.

S   Cost of marking, or laying out all the pattern pieces necessary to make a full size range of a garment or group of garments. Due to advances in technology, much of this work is done on CAD/CAM systems. Generally, a total price is quoted and the designer or manufacturer will divide the cost by the number of pieces being produced to determine the price for one garment. This expense is generally incurred once and will not show up on a reorder.

T   Cost incurred for developing a full size range of products. Often this cost is combined with marking costs.

U   A price given by the manufacturer that is based on a minimum cut. To determine cost per unit, divide the fee by the number of pieces. For example, a $100.00 fee for a 200-piece minimum cut yields 50¢ per garment.

V   Labor costs, including bundling, sewing, pressing, trimming, and inspection. Generally determined as piecework (per-piece) costs.

W   Total labor costs for one garment.

X   Duties based on the shipment. See Y.

Y   Costs based on fiber content, weight of shipment, where it is being manufactured, and where it is shipped. Freight companies and brokers will be able to calculate the cost of freight and advise on the import duties and taxes. These figures are given as a total and must be divided out per piece.

Z   Total labor and shipping costs per garment.

AA   Total cost of the goods to be sold: materials + trims + labor + shipping and duty expenses.

BB   The price that the buyer pays: cost of goods to be sold + markup = wholesale price. Markup is the amount of money added on that allows the designer or manufacturer to cover expenses and make a profit. Markup must also be low enough to keep the price competitive. (Review Chapter 4.)

CC   The suggested price that the retail consumer would pay. This is determined by wholesale price + markup = retail price.

Consider a cost sheet that is filled in and see how the values are determined. This information on this cost sheet gives the design team the following information:

1. On June 12, 19— these were the negotiated prices.
2. This is a woman's two-piece knit style 510 that is to be manufactured in the United States, with the fabric coming from a mill in North Carolina.
3. There are four sizes being made; but out of every six pieces, there will be one small, two mediums, two larges, and one extra large.
4. The bolt of fabric is 60 inches wide. The markers have determined that 10.5 yards are needed for one full scale run to be made. Therefore, $10.5 \div 6 = 1.75$ yards per average piece, based on the size scale.
5. The material is $6.40 per yard, which is the amount per yard that was determined after all the trade, quantity, and seasonal discounts were deducted. Sometimes even the shipping charges are calculated into this price per yard if the shipping charges are high. Therefore, 1.75 yards $\times$ $6.40 = $11.20 per average piece.
6. Also, 0.25 yard of fabric is needed for a collar trim: $0.25 \times $1.25 = $0.31
7. Total cost of materials is $11.51.
8. The trims include one button at $0.04 each, for a total of $0.04, and 0.5 yard of elastic tape at $1.50 per yard, for a total of $0.75. The total cost of the trims is $0.79.
9. Labor expenses are determined as follows:

- *Marking costs*   These charges are usually quoted for minimum cuts. In this case the minimum is a 100-piece cut. For every 100 pieces, the marking charge is $50.00; therefore, the marking expense is $50.00 $\div$ 100 pieces = $.0.50. Some firms write $0.50 on the form, but others prefer to record the total charge because they then stay aware of what the minimum cut has to be. If a larger amount is made, there might be a lower price per piece.
- *Grading costs*   For every 100 pieces the cutting charge is $50.00. ($50.00 $\div$ 100 pieces = $0.50)
- *Cutting costs*   For every 100 pieces, the cutting charge is $72.00. ($72.00 $\div$ 100 pieces = $0.72)
- *Construction costs*   The designer or manufacturer has negotiated a construction cost of $4.48 for each garment. The manufacturing company has presented this offer based on the amount of work in each garment.

10. The total labor costs are $6.20, which is determined by adding $0.50 + $0.50 + $0.72 + $4.48 = $6.20. The total *cost of goods to be sold* is $11.51 + $0.79 + $6.20 = $18.50. This is what you, the designer or wholesaler, will actually pay for each garment. (CGS)

COST SHEET

Date _June 12, 19—_

Style number _510_

Sketch and materials swatch

Description _Woman's two-piece knit_

Fabric mill _North Carolina_

Manufactured _U.S.A._

Size range _6, 8, 10, 12;_

   Size scale _1-2-2-1_

Fabric width _60 in._

Estimated marker yardage _10.5 yds_

Actual marker yardage _____

Materials	Yardage	Price per Yard	Total Amount
Fabric	1.75 yd	$6.40	$11.20
Fabric (collar)	0.25 yd	$1.25	$0.31
Lining	_____	_____	_____
		Total cost of materials	$11.51

Trims	Amounts	Cost Each	Total Amount
Buttons	1 button	$0.04	$0.04
Zippers	none	_____	_____
Belts	none	_____	_____
Other (elastic tape)	$0.50 yd	$1.50/yd	$0.75
		Total cost of trims/findings	$0.79

Labor

Marking costs _$0.50_   Grading costs _$0.50_   Cutting costs _$0.72_

Construction _$4.48 each_

                   Total labor costs _$6.20_

Shipping and Duty Expenses

Import duties _none_   Shipping expenses _none_

              Total shipping and duty costs _none_

Total cost of goods to be sold _$18.50_

Suggested wholesale price _$33.75_

Suggested retail price _$75.00_

Calculating Markup

Once the cost is determined, the designer or wholesaler then has to calculate the selling price—the wholesale price when this garment is sold to a store buyer. It is important to remember that the selling price is a combination of two parts:

1. The actual cost of goods to be sold (CGS), in this case $18.50
2. Markup, the amount of money planned to cover expenses and make a profit

In the wholesale industry the markup is generally 30% to 50% of the total selling price. Let's see how to calculate this value.

The selling price (which is always the total price) is equal to 100%. If the desired markup is 45%, the cost of goods to be sold must be 55%. This means that the final selling price, 100%, is a combination of 45% and 55%.

If you actually put this information in a T-chart, it is very easy to see.

	Dollars	**Percents**
Selling price		100%
Markup		45%
CGS	$18.50	55%

Now simply remember that you are looking at the parts in relationship to the whole, and you know that $18.50 = 55%. If you divide, you will get the selling price:

$$\$18.50 \div 55\% = \$33.64$$

Typically, a designer or manufacturer rounds this amount to the nearest quarter, so let's round up to $33.75. Fill that amount in the following chart and, using some basic math, check your work.

	Dollars	**Percents**
Wholesale cost		
Selling price	$33.75	100%
Markup		45.19%
CGS	$18.50	54.81%

	Dollars	**Percents**
Selling price	$33.75	100%
Markup	$15.25	45.19%
CGS	$18.50	54.81%

Check your work:

$$\$33.75 - \$18.50 = \$15.25$$
$$\$33.75 \times 0.4519 = \$15.25$$
$$\$33.75 \times 0.5481 = \$18.50$$

The $15.25 the designer or wholesaler charges has to be enough to pay the salaries and expenses of the design team and the rest of the operation. In addition, each company is in a business to make a profit, and there must be money left after all the expenses have been paid.

As you can see, the price is now inching its way up, because the wholesale price is just the price at which the designer or wholesaler will sell the merchandise to the store. Now the designer must determine what the suggested retail price will be to the *consumer*. (It is the store buyer's job to make the final determination of the retail price, but a good wholesaler will have a pretty clear idea of what that price will have to be.) The most important thing product designers must ask themselves is this: *How much will the customer pay for this garment?*

Store Markup

In the retailing industry it is expensive to run a company, and sellers are competing for everyone's business. The garment must be quite exciting to command enough markup. There are markup standards in the industry, which can be researched in the FOR reports. These are the financial operating reports prepared by the National Retail Federation. The rule of thumb is that retailers need to earn between 45% and 60% initial markup to be profitable. The same rules apply: Retail markup has to be enough to cover the store's expenses and still make a profit (and yet be at a competitive price so the consumer will buy it, too).

Mathematically, the same concept applies when calculating retail prices. The wholesale or cost price plus a markup determine the actual retail price. And, the retail price is the total, which is always equal to

100%. Let's take a look at a T-chart; it shows the cost at $33.75, since that is what the store paid.

	Dollars	**Percents**
Retail selling price		100%
Markup		
Cost/wholesale	$33.75	

In this case the designer has decided that the sales team will "pitch" a suggested retail price of $75.00 to the buyers during market. Let's see what kind of markup can be earned.

	Dollars	**Percents**
Retail selling price	$75.00	100%
Markup		
Cost/wholesale	$33.75	

Again, you are going to be working with parts of the whole and looking for the relationship, but this time, you are going to work with the numbers. Let's find out what the markup percent really is.

$75.00 − $33.75 = $41.25
$41.25 ÷ $75.00 = 55.0%   (This is the markup the store gets.)
$33.75 ÷ $75.00 = 45.0%

Once again, check your work. You have two parts, the markup and cost, that have to equal the total amount, which is always 100%.

$$\$41.25 + \$33.75 = \$75.00$$
$$55.0\% + 45.0\% = 100\%$$

The math formulas for all markup problems are the same formulas you learned in Chapter 4:

$$\text{Selling price} = 100\%$$
$$\text{Cost\$} \div \text{cost\%} = \text{selling price}$$
$$\text{Markup \$} \div \text{markup\%} = \text{selling price}$$
$$\text{Selling price} \times \text{markup\%} = \text{markup \$}$$
$$\text{Selling price} \times \text{cost\%} = \text{cost \$}$$

Once you have made all the calculations, then you finally know if the numbers will support the illustrations, the colors, the fabrics, and the shipping plans.

Costing and Pricing: Wholesale versus Private Label

You just costed out a two-piece knit for $18.50 that will sell in the department store for $75.00. The designers must sell their products to the stores, and the stores in turn sell the products to the consumer. There are lots of middle people in that process—fashion shows are held, major publicity and public relations events are staged, and dozens of people along the selling–customer service ladder all have to be paid. The designer still is going to make only $15.25 in gross margin or markup, and that has to be enough to cover the company expenses and make a profit.

But, suppose you were a product-development specialist for a major department store. You would go to the same fabric shows and probably visit the same factories to make your products. The difference now is that you don't have to sell it to a store. You are making it for your own store. You will incur some advertising expenses to promote the product, but consider this:

	Dollars	**Percents**
Retail selling price	$75.00	100%
Markup		
CGS	$18.50	

If you marked the two-piece knit the same price as the designer, look at how much margin or markup you could make:

$$\$75.00 - \$18.50 = \$56.50$$

That's a lot more than the designer makes.

If you fill in the numbers and percentages on the T-chart, it will look like this:

	Dollars	**Percents**
Retail selling price	$75.00	100.0%
Markup	$56.50	75.3%
CGS/wholesale	$18.50	24.7%

Remember, you did not change the cost of the goods sold at all. However, you must also realize that customers will usually pay more for a well-recognized national brand. Private labels tend not to command as much loyalty. So what if you *reduce* your selling price? You are still giving the consumer the same product but at a sharper price. *That* might get the customer to buy.

Let's see what your profits will look like if you mark the two-piece knit at $60.00.

	Dollars	**Percents**
Retail selling price	$60.00	100.0%
Markup	$41.50	69.17%
CGS/wholesale	$18.50	30.83%

Note that $60.00 − $18.50 = $41.50 and then $41.50 ÷ $60.00 = 69.17%. Just follow the math process you have used before.

At $60.00 you are still providing a nice margin for your company (69.17%) to cover expenses and make a profit, but suddenly you are cutting into the designer's business and giving the consumer a much better price. The challenge a product-development or private-label buyer faces is to build strong and consistent name recognition. Private-label stores such as Banana Republic, Express, Structure, Laura Ashley, and Coach have reached this goal, and now the department stores are joining them. Consumers are seeking out the Charter Club, INC, Field Gear, and Arizona Jeans brands, and the department store organizations are pleased.

■ Problems

1. Use the given information.

   a. Determine the cost of goods to be sold. (CGS)

Spring bridesmaid dress

Date: August 8, 19—

Style: 875

Sizes: 6, 8, 10, 12, 14

Size scale: 1–2–3–3–3

Fabric mill: Taiwan

Manufactured: Hong Kong

Fabric width: 36 in.

Estimated marker yardage: 39 yds

Material: Floral print chintz, $3\frac{1}{4}$ yds at $2.48 per yard (This yardage for one garment is determined by estimating the total yardage for a one-dozen run and then dividing it by 12: 39 yds ÷ 12 pieces = 3.25 yds for one garment, which is entered on the cost sheet.)

Pellon lining: $\frac{3}{4}$ yd at $1.29 per yard

Trim: 4 yds at $0.67 per yard

Buttons: One decorative button, $0.35 each

Marking and grading costs: $285.00 based on a 500-piece minimum order (combined cost) (That means $285.00 ÷ 500 is the cost for one piece.)

Cutting costs: $175.00 based on a 500-piece minimum order

Construction costs: $84.00 per dozen (This is for all 12; what is it for one?)

Duties/shipping: $1.12 each garment

   b. Determine the wholesale selling price using a 60% markup. Fill in the T-chart grid.

	Dollars	**Percents**
Wholesale price		100%
Markup		60%
CGS		40%

COST SHEET

Date _____

Style number _____

Sketch and materials swatch

Description _____

Fabric mill _____

Manufactured _____

Size range _____

Size scale _____

Fabric width _____

Estimated marker yardage _____

Actual marker yardage _____

Materials	Yardage	Price per Yard	Total Amount
Fabric	_____	_____	_____
Fabric	_____	_____	_____
Lining	_____	_____	_____

Total cost of materials _____

Trims	Amounts	Cost Each	Total Amount
Buttons	_____	_____	_____
Zippers	_____	_____	_____
Belts	_____	_____	_____
Other	_____	_____	_____

Total cost of trims/findings _____

Labor

Marking costs _____ Grading costs _____ Cutting costs _____

Construction _____

Total labor costs _____

Shipping and Duty Expenses

Import duties _____ Shipping expenses _____

Total shipping and duty costs _____

Total cost of goods to be sold _____

Suggested wholesale price _____

Suggested retail price _____

Remember, go back and look at your formulas. Selling percent is always equal to 100%; in this case, since you want the markup percent to be 60%, the cost percent will be 40%. Simply divide the cost of goods sold by the cost percent and you will arrive at the wholesale selling price.

c. Determine the retail selling price. In this chapter we rounded our wholesale price to the nearest quarter ($0.00, $0.25, $0.50 or $0.75). Based on the wholesale price you reached, round the wholesale price accordingly and use this value in the following T-chart. Then, applying a 50% retail markup, determine the retail selling price for this garment.

	Dollars	**Percents**
Retail selling price		100%
Markup%		50%
Cost/wholesale price		50%

d. Do you think this will be a strong selling item? Or do you need to go back and look for a better price on some element of the cost sheet?

2. Use the given information.

    a. Determine the cost of goods to be sold. (CGS)

Men's three-button placket-front cotton knit shirt

Date: December 5, 19—

Style: 454

Sizes: S, M, L, XL

Size scale: 1–4–5–2

Fabric mill: North Carolina

Manufactured: Miami, Florida

Fabric width: 60 in.

Estimated marker yardage: 24 yds for 12 pieces (One shirt will be 2 yds, the number to enter in the yardage column on the cost sheet.)

Material: Cotton knit jersey at 1.60/yd

Buttons: 3 buttons at $0.03 each

Marking and grading costs: $185.00 based on a 500-piece minimum (combined cost)

Cutting costs: $175.00 based on a 500-piece minimum

Construction costs: $42.00 per dozen

Shipping: FOB store

    b. Determine the wholesale selling price using a 40% markup. Fill in the T-chart.

	Dollars	**Percents**
Wholesale price		100%
Markup		40%
CGS		60%

Remember, go back and look at your formulas. Selling percent is always equal to 100%; in this case, since you want the markup percent to be 40%, the cost percent will be 60%. Simply divide the cost of goods sold by the cost percent and you will arrive at the wholesale selling price.

COST SHEET

Date _____

Style number _____

Sketch and materials swatch

Description _____

Fabric mill _____

Manufactured _____

Size range _____

Size scale _____

Fabric width _____

Estimated marker yardage _____

Actual marker yardage _____

Materials	Yardage	Price per Yard	Total Amount
Fabric	_____	_____	_____
Fabric	_____	_____	_____
Lining	_____	_____	_____
Total cost of materials			_____

Trims	Amounts	Cost Each	Total Amount
Buttons	_____	_____	_____
Zippers	_____	_____	_____
Belts	_____	_____	_____
Other	_____	_____	_____
Total cost of trims/findings			_____

Labor

Marking costs _____ Grading costs _____ Cutting costs _____

Construction _____

Total labor costs _____ _____

Shipping and Duty Expenses

Import duties _____ Shipping expenses _____

Total shipping and duty costs _____

Total cost of goods to be sold _____

Suggested wholesale price _____

Suggested retail price _____

c. Determine the retail selling price. In this chapter we rounded our wholesale price to the nearest quarter ($0.00, $0.25, $0.50 or $0.75) when setting retail prices. Based on the wholesale price you reached, round the wholesale price accordingly and use this value in the following T-chart. Then, applying a 55% retail markup, determine the retail selling price for this garment.

	Dollars	Percents
Retail selling price		100%
Markup		55%
Cost/wholesale price		45%

d. Do you think this will be a strong selling item? Or do you need to go back and look for a better price on some element of the cost sheet?

e. If you are working in a resident buying office in the product-development division or if you are working with a major chain designing garments that will carry a private label, calculate the retail selling price, earning exactly the same dollars in retail markup as you calculated in part (c).

	Dollars	Percents
Retail selling price		100%
Markup		
CGS		

Notice how much lower the retail selling price is, even though your company is earning the same retail markup dollars and the cost of the goods to be sold is the same.

f. If you were a designer or manufacturer, what would you do with your product line to compete with this price point on the private-label goods?

3. Use the given information.

  a. Determine the cost of goods to be sold.

  Missy's sportswear: Jogging shorts
  Date: May 14, 19—
  Style: 1890
  Sizes: S, M, L
  Size Scale   2–4–6
  Fabric mill: Honduras
  Manufactured: Honduras
  Fabric width: 45 in.
  Estimated marker yardage: 7.5 yds for 12 pieces, 0.625 yd for one pair (That number will go in the yardage column on the cost sheet.)
  Material: Nylon at $1.45/yd
  Elastic: $\frac{3}{4}$ yd at $0.65/yd
  Marking and grading costs: $50.00 based on a 200-piece minimum (combined cost)
  Cutting costs: $90.00 based on a 200-piece minimum
  Construction costs: $1.50 each
  Shipping and duties: $0.82 per each garment

  b. Determine the wholesale selling price using a 30% markup. Fill in the T-chart.

	Dollars	**Percents**
Wholesale price		100%
Markup		30%
CGS		70%

  Remember, go back and look at your formulas. Selling percent is always equal to 100%; in this case, since you want the markup percent to be 30%, the cost percent will be 70%. Simply divide the cost of goods to be sold by the cost percent and you will arrive at the wholesale selling price.

  c. Determine the retail selling price. Based on the wholesale price you reached, round the wholesale price accordingly and use

COST SHEET

Date _____

Description _____

Style number _____

Fabric mill _____

Sketch and materials swatch

Manufactured _____

Size range _____

Size scale _____

Fabric width _____

Estimated marker yardage _____

Actual marker yardage _____

Materials	Yardage	Price per Yard	Total Amount
Fabric	_____	_____	_____
Fabric	_____	_____	_____
Lining	_____	_____	_____

Total cost of materials _____

Trims	Amounts	Cost Each	Total Amount
Buttons	_____	_____	_____
Zippers	_____	_____	_____
Belts	_____	_____	_____
Other	_____	_____	_____

Total cost of trims/findings _____

Labor

Marking costs _____   Grading costs _____   Cutting costs _____

Construction _____

Total labor costs _____

Shipping and Duty Expenses

Import duties _____   Shipping expenses _____

Total shipping and duty costs _____

Total cost of goods to be sold _____

Suggested wholesale price _____

Suggested retail price _____

this price in the following T-chart. Then, applying a 45% retail markup, determine the retail selling price for this garment.

	Dollars	**Percents**
Retail selling price		100%
Markup		45%
Cost/wholesale price		55%

d. Do you think this will be a strong selling item, or do you need to go back and look for a better price on some element of the cost sheet?

e. If you are working in a resident buying office in the product-development division or if you are working with a major chain designing garments that will carry a private label, calculate the retail selling price, earning exactly the same dollars in retail markup as you calculated in part (c).

	Dollars	**Percents**
Retail selling price		100%
Markup		
CGS		

Notice how much lower the retail selling price is, even though your company is earning the same retail markup dollars and the cost of the goods sold is the same.

f. If you are a designer or manufacturer, what would you do with your product line to compete with this price point on the private-label goods?

4. As a merchandising designer for a sourcing firm, you go on a buying trip to your overseas company-owned factory and design a woman's sweater that carries a cost-of-goods-sold price of $16.50. This price includes duty and freight charges. In the market, the customer is

currently paying $85.00 for this style sweater. Your goal is to bring the sweater to the United States and sell it at a wholesale price of $41.25 to the department stores. From this price, the retailer marks the merchandise to $79.99, which still undercuts the current market price. Complete the following T-chart grids, identifying the cost dollars and retail dollars along with the markups in both dollars and percents.

	Dollars	Percents		Dollars	Percents
Cost/wholesale		100%	Retail		100%
Markup			Markup		
CGS			Cost		

Once you make those calculations, identify the total markup dollars achieved on this product for the company.

5. Boy's rain jackets, sizes 8–20, have been designed for fall by an Asian manufacturer. The cost of goods is $48.00 a dozen. This cost includes all duties and import fees. As a merchandising designer for a product-development firm, you need to determine a wholesale price and provide a suggested retail price that will be competitive for back-to-school sales. Your firm generally works with a 45% markup to determine wholesale and suggests an additional 50% markup to determine the retail selling price. Complete both T-charts to determine the wholesale, or cost, price and the retail selling price.

	Dollars	Percents		Dollars	Percents
Cost/wholesale		100%	Retail		100%
Markup			Markup		
CGS			Cost		

Once you have determined an actual retail selling price, decide what you feel would be a strong retail promotional price for this type of garment. Explain.

Workbook Aids:
Formulas and Answers

Basic Business Formulas

Helpful Hints

Total amount × percent = specific amount

Specific amount ÷ total amount = percent

Specific amount ÷ percent = total amount

Difference ÷ original amount = percent of increase (or decrease)

Basic Markup Formulas

Cost + markup = retail (dollars or percents)

Retail − cost = markup (dollars or percents)

Retail − markup = cost (dollars or percents)

Retail × cost% = cost$

Retail × markup% = retail markup$

Retail markup$ ÷ retail$ = retail markup%

Cost$ ÷ retail$ = cost%

Cost$ ÷ cost% = retail$

Retail markup$ ÷ retail markup% = retail$

Selling price = 100%

100% − markup% = cost%

Analyzing and Evaluating Six-Month Merchandising Plans

Monthly sales ÷ total sales = monthly sales percent

All 6-month sales percentages = 100%

BOM stock ÷ monthly sales = monthly stock-to-sales ratio (SSR)

Monthly markdowns ÷ monthly sales = monthly markdown%

Total markdowns ÷ total season sales = overall markdown%

(6 BOM + 1 EOS) ÷ 7 = average stock for a season

LY EOS ÷ LY total sales = EOS% (used when planning with LY figures)

Net sales ÷ average stock = turnover rate

Monthly sales + EOM stock +
monthly markdowns − BOM = retail purchases

Retail purchases × cost% = cost value of purchases

Developing a Six-Month Merchandising Plan

Sales volume × monthly sales% = monthly sales volume

Monthly planned sales volume × SSR = monthly planned BOM

Net sales × overall markdown% = total markdown$

Monthly markdown$ (allocated) ÷ monthly sales = monthly markdown%

(6 BOM + 1 EOS) ÷ 7 = average stock for a season

TY planned sales × EOS percent = TY planned EOS stock
(used when planning with LY figures)

Net sales ÷ average stock = turnover rate

Monthly sales + EOM stock + monthly markdowns − BOM = retail purchases

Retail purchases × cost% = cost value of purchases

Profitability

Gross sales − customer returns and allowances = net sales

Net sales − cost of goods sold = gross margin

Gross margin − operating expenses = profit (loss)

Length × width = square feet

Net sales ÷ square feet = sales per square foot

Cost of Goods Sold (CGS)/Manufacturer's Cost Sheet

Material costs + trim cost + labor + shipping and duty expenses
= cost of goods to be sold (CGS)

Answers

■ Chapter 2

1. $669.20

3. 924.32   Check your work and round if necessary: 924.32 ÷ 106% = 872

5. 2.96%   Check your work and round if necessary: $8,100 × 2.96% = 240

7. 4.98%   Check your work and round if necessary:
   $1,050 × 4.976% = $52.25. Round 4.976% to 4.98%.

9. 2,007.50   Check your work and round if necessary: $2,007.50 ÷ 55% = $3,650

11. $10,824.05   Check your work and round if necessary:
    $10,824.05 × 39.5% = $4,275.50

13. 400   Check your work. Is this answer logical?

15. 33.87%   Check your work and round if necessary:
    $185.95 × 33.87% = $62.98

17. 10% increase   Check your work: $2,500.00 + 10% × $2,500 = $2,750

19. 26% increase   Check your work: $250,000 + 26% × $250,000 = $315,000

21. 2 x 12 = 24; 24 + 9 = 33

23. 8, 30, 39, 64, and 216

25. 224, 18, 57, 136, 110, 90, 180, 242

27. 6 x 12 = 72 pens originally on sale
$6 - 2\frac{1}{2} = 3\frac{1}{2}$ doz. were sold
$2\frac{1}{2}$ doz. = 2 x 12 + 6 = 30 pens left
$3\frac{1}{2}$ doz. = 36 + 6 = 42; 42 x $8.99 = $377.58

29. 60 x 0.75 = 45 sold; use your calculator: 60 ☒ 75 ▨   45
60 − 45 = 15 left; use your calculator: 60 ⊟ 75 ▨   15

31. $15.75 ÷ 3.50 = 4.5; 4.5 x 2 = 9

33. 8 + 6 + 11 = 25; 25 x 12 = 300; 300 + 4 + 3 + 8 = 315

35. $426,000.00 x 12% = $51,120.00; $51,120.00 x 6% = $3,067.20

37. 20 x 12 + 8 = 248; 248 x 25% = 62

39. $1,560.00 − $1,200.00 = $360.00

$360.00 ÷ $7,200.00 = 5%

Check your work: 7200 [×] 5[%]   360

$1,200.00 + $360.00 = $1,560.00

41. 38 ÷ 95% = 40 sweaters

43. $364,000.00 − $260,000.00 = $104,000.00

$104,00.00 ÷ $260,000.00 = 40% increase

Check your work with your calculator: 260000 [+] 40[%]   364000

45. Use your calculator: 415 [−] 20[%]   332

47. Use your calculator: 16,500 [÷] 4[%]   412,500

49. 24 × $28.00 = $672.00

19 × $32.00 = $608.00

26 × $15.00 = $390.00

4 × 12 + 6 = 54; 54 × $8.00 = $432.00

$672.00 + $608.00 + $390.00 + $432.00 = $2,102.00

Use your calculator: 2102 [÷] 152,000 [%]   1.382   or 1.382%

51. 100% + 28% = 128%-this year's planned sales

Use your calculator: 615000 [÷] 128[%]   480,468.75

Check your work with your calculator: 480,468.75 [+] 28% [=]   615,000

53. Use Helpful Hint 2:

Janelle   10%   Reuben   7.14%   Omar   11.33%

Then simply calculate sales productivity as follows:

Total sales for each employee ÷ number of hours worked = sales productivity

Janelle $131.25   Reuben $187.50   Omar $74.38

55. Use your calculator: 550 ⊟ 60 ⏛ | 220 |

56. Use your calculator: 325000 ÷ 132 ⏛ | 246212.12 |
    Check your work: 246212.12 ⊞ 32 ⏛ | 324999.99 |

59. Use your calculator: 750000 ⊟ 14 ⏛ | 645000 |

61. Use your calculator: 1.88 ⊠ 75 ⏛ 1.41 ÷ 2.5 ⏛ | 56.4 |
    Check your work: 2.5 ⊠ 56.4 ⏛ 1.41

63. 30000000 ⊠ 1.9 ⏛ | 570000 | LY net sales for hosiery
    $570,000 − $5,700 = $564,300   Sales 2 years ago
    30000000 ⊠ 2.1 ⏛ | 630000 | LY net sales for handbags
    $630,000 + $14,000 = $644,000   Sales 2 years ago

65. Use Helpful Hint 3 and your calculator: 62000 ÷ 118 ⏛ | 52542.37 |
    Last year sales volume
    Use Helpful Hint 4 and your calculator: 62000 ⊟ 12000 ÷
    12,000 ⏛ | 416.66666 |

■ Chapter 3

1. April 10, April 10
   April 18, April 18
   Nov. 10, Nov. 10
   Aug. 7, Aug. 6
   Sept. 30, Sept. 30
   Feb. 10, Feb. 10
   Oct. 30, Oct. 30
   June 15, June 14
   April 30, May 1
   Jan. 21, Jan. 20
   Dec. 12, Dec. 12

3. 1295.00 ⊟ 30⅟% 906.50 ⊟ 15% = ▓770.525▓ or $770.53

(Remember, with series discounts, deduct each discount individually from the previous amount.)
January 22 pay: 770.53 ⊟ 6% ▓724.3▓ or $724.30

5. February 28: $419.78
March 10: $432.77

Discussion: It is always best to earn all discounts if possible.

7. $3,937.50

9. The store buyer

11. a. $107.53
b. $408.00

(Remember, with series discounts, deduct each discount individually from the previous amount.)

13. $60.00; cash discount is only on billed cost. Freight charges are always added on last.

15. $3,920, with 70 days to pay to earn the 2% cash discount

17. 76 – 40 = 36 dozen
40 ⊠ 8.5 ⊟ ▓340▓
36 ⊠ 8.5 ⊟ 5% = 290.7

Net cost is $630.70.

19. $1,041.84

21. 16 × $24.00 = $384.00
36 × $12.00 = $432.00
24 × $10.00 = $240.00
1056 ⊟ 35 % ⊟ 15 % = ▨ 583.44 ▨ or $583.44

The store did not have to pay for the transportation charges. The store did not earn an additional 3% cash discount because the bill was due on September 10 and was not paid until September 12.

23. 14 × 12 = 168; 168 × $4.50 = $756.00
16 × $65.00 = $1040.00
20 × 108.00 = $2160.00

The total invoice is $3,956.00.
3956 ⊟ 5 % ▨ 3758.2 ▨ or $3,758.20
3758.20 ⊟ 8 % ▨ 3457.544 ▨ or $3,457.54 if paid on or before September 22

24. Linen company B is the better buy if the bill is paid on time.
Linen company A: 10000 ⊟ 4 % – 2 % ▨ 9408 ▨ + shipping charges.
Linen company B: 10000 ⊟ 8 % ▨ 9200 ▨ no shipping charges.

25. 6 × 12 = 72 × 20 = 1440; 12½ × 12 = 150 × 12.50 = 1875;
8 × 12 = 96 × 5 = 480; 1440 + 1875 + 480 = 3795. Use your calculator: 3795 ⊟ 8 % ▨ 3491.4 ▨ ⊞ 83.76 ⊟ ▨ 3575.16 ▨

■ Chapter 4

1.

Retail Dollars	Cost Dollars	Retail Markup Dollars	Retail Markup Percent
$ 68.00	$ 23.00	$ 45.00	66.18%
$465.50	$169.99	$295.51	63.48%
$440.00	$ 110.00	$330.00	75%
$230.77	$120.00	$ 110.77	48%
$ 14.50	$ 7.83	$ 6.67	46%
$ 178.57	$100.00	$ 78.57	44%
$ 5.43	$ 2.50	$ 2.93	54%
$ 3.00	$ 1.50	$ 1.50	50%

3. $\$6.00 \times 33.3\% = \$2.00$
   $\$6.00 - \$2.00 = \$4.00$

5. $\$150.00 - \$60.00 = \$90.00$
   $\$90.00 \div \$150.00 = 60\%$
   $100\% - 60\% = 40\%$

7. $100\% - 49\% = 51\%$
   $\$12.75 \div 51\% = \$25.00$

9. $\$132.00 \div 12 = \$11.00$ each
   $\$22.00 - \$11.00 = \$11.00$
   $\$11.00 \div \$22.00 = 50\%$

11. $\$51.50 - \$23.00 = \$28.50$
   $\$28.50 \div \$51.50 = 55.34\%$

13. $\$9.00 \div 12 = \$0.75$ each
   $\$2.00 - \$0.75 = \$1.25$
   $\$1.25 \div \$2.00 = 62.5\%$
   $100\% - 62.5\% = 37.5\%$

15. $100\% - 40\% = 60\%$
   $\$90.00 \div 60\% = \$150.00$

17. $100\% - 58\% = 42\%$
   $\$3.95 \times 42\% = \$1.66$

19. $84.00 ÷ 60\% = \$140$

21.

Units	Cost Each	Cost Total	Retail Each	Retail Total
12 doz./144 pcs	$12.60/doz.	$151.20	$1.89	$ 272.16
27 doz./324 pcs	$13.20/doz.	$356.40	$1.99	$ 644.76
11 doz./132 pcs	$ 9.00/doz.	$ 99.00	$1.29	$ 170.28
Total 50 doz./600 pcs		$606.60		$1,087.20

	Dollars	Percents
R	$1,087.20	100%
MU	$ 480.60	44.2%
C	$ 606.60	55.8%

23.

Units	Cost Each	Cost Total	Retail Each	Retail Total
30	$54.00	$1,620.00	$108.00	$3,240.00
24	$28.00	$ 672.00	$ 60.00	$1,440.00
18	$16.00	$ 288.00	$ 32.00	$ 576.00
Total 72 pcs		$2,580.00		$5,256.00

	Dollars	Percents
R	$5,256	100%
MU	$2,676	50.91%
C	$2,580	49.09%

25.

	Dollars	Percents
R	$8.39	100%
MU	$4.39	52.3%
C	$4.00	47.7%

	Dollars	Percents
R	$120.00	100%
MU	$ 67.44	56.2%
C	$ 52.56	43.8%

Units	Cost Each	Cost Total	Retail Each	Retail Total
200	$36.00	$ 7,200.00	$ 80.00	$ 16,000.00
150	$ 4.00	$ 600.00	$ 8.39	$ 1,258.50
350	$52.56	$18,396.00	$120.00	$42,000.00
Total 700 pcs		$26,196.00		$59,258.50

	Dollars	Percents
R	$59,258.50	100%
MU	$33,062.50	55.79%
C	$26, 196.00	44.21%

27.

	Dollars	Percents
R	$18.95	100%
MU	$ 9.95	52.5%
C	$ 9.00	47.5%

($108.00 ÷ 12 = $9.00 each)

	Dollars	Percents
R	$28.00	100%
MU	$15.00	53.57%
C	$13.09	46.43%

	Dollars	Percents
R	$36.00	100%
MU	$19.80	55%
C	$16.20	45%

Units	Cost Each	Cost Total	Retail Each	Retail Total
180	$ 9.00	$1,620.00	$18.95	$ 3,411.00
200	$13.00	$2,600.00	$28.00	$ 5,600.00
280	$16.20	$4,536.00	$36.00	$10,080.00
Total 660 pcs		$8,756.00		$19,091.00

	Dollars	Percents
R	$19,091	100%
MU	$10,335	54.14%
C	$ 8,756	45.86%

■ Chapter 5

1. *Plans:*

$$48 \times 12 = 576 \text{ pcs}$$
$$576 \times \$25.00 \text{ each} = \$14,400 \text{ total retail}$$
$$100\% - 48\% = 52\%$$
$$\$14,400 \times 52\% = \$7,488 \text{ total cost}$$

On order:

$$26 \times 12 = 312 \text{ pieces}$$
$$312 \times \$12.50 \text{ each} = \$3,900 \text{ cost dollars spent}$$

Open to buy:

$$\$7,488 - \$3,900 = \$3,588 \text{ remaining cost dollars}$$
$$\$3,588 \div 22 \text{ dozen (264 pieces)} = \$13.59 \text{ each}$$

The most the buyer can pay is $13.59 each.

3. *On order:*

4 dozen (48 pieces); cost = $40.00; retail = $84.00

6 dozen (72 pieces); cost = $90.00; retail = $216.00

6 dozen (72 pieces); cost = $72.00

Plan:

$$\text{Total cost dollars} = \$202.00$$
$$100\% - 52\% = 48\%$$
$$\$202.00 \div 48\% = \$420.83 \text{ total planned retail}$$
$$\$420.83 - \$84.00 - \$216.00 = \$120.83$$
$$\$120.83 \div 72 \text{ pieces} = \$1.68$$

The buyer would mark the note cards either $1.65 or $1.75 to keep the same type of price line as the other note cards.

5. *On order:*

12 dozen (144 pieces); cost = $36.00 per dozen; total cost = $432.00; total retail = $720.00

20 dozen (240 pieces); cost = $4.75 each; total cost = $1,140.00; overall cost $1,572.00

Plan:

$$100\% - 43\% = 57\%$$
$$\$1,572 \div 57\% = \$2,757.89$$
$$\$2,757.89 - \$720.00 = \$2,037.89$$
$$\$2,037.89 \div 240 \text{ pieces} = \$8.49 \text{ each}$$

7.
$$100\% - 52\% = 48\%$$
$$\$160,000 \times 48\% = \$76,800$$
$$\$160,000 - \$106,000 = \$54,000$$
$$\$76,800 - \$54,000 = \$22,800$$

	Dollars	Percents
R	$54,000	100%
MU	$31,200	57.78%
C	$22,800	42.22%

9. 30 dozen (360 pieces) × $96.00 = $2,880.00
40 dozen (480 pieces) × $72.00 = $2,880.00
$2,880.00 + $2,880.00 = $5,760.00 (total cost)
$5,760.00 ÷ 48% = $12,000.00 (total retail)
360 pieces × $16.00 = $5,760.00
$12,000.00 − $5,760.00 = $6,240.00
$6,240.00 ÷ 480 pieces = $13.00 each

11.   100% − 52% = 48%
100 × $18.00 = $1,800.00
150 × $15.00 = $2,250.00
250 × $14.00 = $3,500.00
300 × $17.00 = $5,100.00
Total cost = $12,650
$12,650.00 − 10% × $12,650 = $11,385.00; $11,385.00 ÷ 48%
= $23,718.25
$23,718.25 ÷ 800 pieces = $29.65 each

13.   100% − 43% = 57%
165 + 200 = 365 pcs
165 × $25.00 = $4,125.00
200 × $28.00 = $5,600.00
Total cost = $9,725
$9,725 ÷ 57% = $17,061.40
$17,061.40 ÷ 365 pcs = $46.74 each

The buyer will probably run a $46.99 promotion.

15.   100% − 52.5% = 47.5%
275 + 475 = 750 pcs
$40.00 ÷ 12.00 = $3.33
275 × $3.33 = $915.75
475 × $3.25 = $1,543.75
Total cost = $2,459.50
$2,459.50 ÷ 47.5% = $5,177.89
$5,177.89 ÷ 750 pcs = $6.90 each

17. $16 \times 12 = 192$ pcs

Initial markup:

$192 \times \$0.75 = \$144.00$ (total cost)
$192 \times \$1.25 = \$240.00$ (total retail)

Sales:

107 pairs $\times$ \$1.25 each $=$ \$133.75
32 pairs $\times$ \$0.95 each $=$ \$30.40
53 pairs $\times$ \$1.75 each $=$ \$92.75
Actual retail dollars earned $=$ \$256.90

	Initial Markup				Maintained Markup	
	Dollars	**Percents**			**Dollars**	**Percents**
R	\$240.00	100%		R	\$256.90	100%
MU	\$ 96.00	40%		MU	\$112.90	43.95%
C	\$144.00	60%		C	\$144.00	56.05%

19. *Initial markup:*

$200 \times \$7.00 = \$1,400.00; \$1,400.00 - 10\% \times \$1,400.00$
$\quad\quad\quad\quad = \$1,260.00; \$1,260.00 - 15\% \times \$1,260.00$
$\quad\quad\quad\quad = \$1,071.00$ (total billed cost)
$200 \times \$12.99 = \$2,598.00$ (total retail)

Sales:

142 pairs $\times$ \$12.99 each $=$ \$1,844.58
58 pairs $\times$ \$18.00 each $=$ \$1,044.00
Actual retail dollars earned $=$ \$2,888.58

	Initial Markup				Maintained Markup	
	Dollars	**Percents**			**Dollars**	**Percents**
R	\$2,598.00	100%		R	\$2,888.58	100%
MU	\$1,527.00	58.78%		MU	\$1,817.58	62.92%
C	\$1,071.00	41.22%		C	\$1,071.00	37.08%

■ Chapter 6

1. $100.00 − 30% × $100.00 = $70.00 new selling price
   $30.00 × 60 sweaters = $1,800.00 markdown dollars

3. $125.00 − 20% × $125.00 = $100.00

5. 62 pieces; $40.00 to $32.99 = difference = $7.01; 62 × $7.01 = $434.62
   26 pieces; $32.99 to $24.99 = difference = $8.00; 26 × $8.00 = $208.00
   Total markdown dollars = $642.62
        $40.00 − $24.99 = $15.01
        $15.01 ÷ $40.00 = 37.5%

7. $999.00 − $679.00 = $320.00
   $320.00 × 36 pieces = $11,520   (total markdown dollars taken)
   $320.00 ÷ $999.00 = 32.03%

9. $175.00 − $124.99 = $50.01; $50.01 × 100 pieces
                   = $5,001.00   (original markdown)
   $124.99 − $87.99 = $37.00; $37.00 ×64 pieces
                   = $2,368.00   (second markdown)

   Total markdown dollars: $7,369.00

11. $25.00 × 20% = $5.00
    162 pcs × $5.00 = $810.00   (total markdown dollars)

    Sales: 162 × $20.00 = $3,240.00
               38 × $25.00 = $950.00
    Total sales: $4,190.00

13. $68.00 − 20% × $68.00 = $54.40
          $68.00 − $54.40 = $13.60

15.

	New Price	Markdown$	Total MD$
Lightning Blades	$164.25	$54.75	$2,080.50
Turbo Trick Blades	$201.75	$67.25	$3,093.50
MM Hockey Blades	$ 89.25	$29.75	$ 505.75
TR Blades	$134.25	$44.75	$2,774.50
Total markdown			$8,454.25

17.

New Price	Markdown$ Difference	Total MD$
$47.99	$17.01	$ 391.23
$32.99	$12.01	$ 444.37
$29.99	$10.01	$ 520.52
$55.99	$19.01	$ 323.17
Total markdown		$1,679.29

19. $40 \times \$12.50 = \$500.00$ total markdown$

$160 \times \$25.00$ each $= \$4,000.00$

$40 \times \$12.50$ each $= \$500.00$

Total sales volume $= \$4,500.00$

21. Net sales on merchandise sold: $50 \times \$15.00 = \$750$

$30 \times \$20.00 = \$600$

$24 \times \$ 8.00 = \$192$

Total net sales $= \$1,542.00$

Actual markdowns taken: $50 \times \$5.00 = \$250$

$30 \times \$5.00 = \$150$

$24 \times \$4.00 = \$96$

Total markdowns $= \$496$

Monthly markdown percent: $\$115,000 - 7.3\% =$ net sales

Net sales $= \$106,605.00$

Total markdown$\% = 0.47\%$

23. $45.00 − 35\% = \$29.25$
$36.00 − 35\% = \$23.40$

Sales:

$29.25 \times 36 = \$1,053.00$
$23.40 \times 23 = \$538.20$
Total sales volume is \$1,591.20.

Markdowns:

$$\$45.00 − \$29.25 = \$15.75; \ \$15.75 \times 36$$
$$= \$567.00 \quad (\text{markdowns})$$
$$\$36.00 − \$23.40 = \$12.60; \ \$12.60 \times 23$$
$$= \$289.80 \quad (\text{markdowns})$$
Total markdowns are \$856.80.

25. *Initial purchase:*

$$144 + 96 = 240 \text{ pairs of sandals}$$
$$240 \times \$16.00 = \$3,840.00 \quad (\text{total cost})$$
$$144 \times \$36.00 = \$5,184.00$$
$$96 \times \$32.00 = \$3,072.00$$

Sales results:

$24 \times \$36.00 = \$864.00 \quad 32 \times \$32.00 = \$1,024.00$

Markdowns and new price:

$36.00 − 25\% = \$27.00 \quad \$32.00 − 25\% \times \$32.00 = \$24.00$

Sales results:

$84 \times \$27.00 = \$2,268.00 \quad 46 \times \$24.00 = \$1,104.00$

Markdowns and new price:

$27.00 − \$17.99 = \$9.01 \quad \$24.00 − \$14.99 = \$9.01$

Sales results:

$36 \times \$17.99 = \$647.64 \quad 18 \times \$14.99 = \$269.82$

Total sales volume was \$6,177.46.

	Initial Markup Dollars	Percents			Maintained Markup Dollars	Percents
R=	$8,256	100%		R=	$6,177.46	100%
MU	$4,416	53.49%		MU	$2,337.46	37.84%
C	$3,840	46.51%		C	$3,840.00	62.16%

Total markdown: $8,256.00 – $6,177.46 = $2,078.54

■ Chapter 7

1. Spring and fall

3. Last year, LY; this year, TY

5. Some factors include but are not limited to weather, new-store openings, store closings, and change in traffic patterns.

7. For example: If the Easter holiday is early in a year—i.e., late March—stock of spring merchandise has to peak in late February, with more aggressive promotions to move out the winter merchandise. If Easter falls mid to late April, buyers have more time to clear out winter merchandise and peak stocks in March.

9. The buyer needs to know what the sales volume was overall and look at how much money was earned each month. From that point all comparisons can be made to the stock figures.

11. The stock on hand was sold and replaced twice during the 6-month period.

13. The merchandise appears "old-looking" to the consumer; it begins to get shopworn and often results in higher markdowns than were planned.

15. Yes. If merchandise is selling too quickly, size assortments break down, as do choices for the customer. Ultimately, the consumer is frustrated by not being able to find pieces or sizes, and they will shop in a store where there is more depth in the assortments.

17. To see what the flow of merchandise is each month. Too much merchandise can overload a department and the consumer. Too little means the consumer doesn't have any new choices.

19. 100% − markup% = cost%
    Retail$ × cost% = cost$

■ Chapter 8

1. Sales% each month:
   Monthly sales ÷ total sales

   18%, 12%, 14%, 22%, 24%, 10%

   Stock-to-sales ratio:
   BOM stock ÷ monthly sales

   2.2, 2.6, 2.6, 2.3, 1.9, 2.7

   Markdown% per month:
   Monthly markdowns ÷ monthly sales

   4.2%, 5.4%, 8.2%, 6.7%, 9.4%, 17.25%

   Total markdowns ÷ total sales

   8.02%

   Monthly purchases at retail:

   $16,777 + $29,914 + $46,612 + $29,990 + $11,795 + $23,750 = $158,838

   Monthly sales + EOM stock
   + monthly markdowns − BOM = purchases

   Monthly purchases at cost:
   100% − MU% = cost%
   Retail$ × cost% = cost$

   $8,724.04 + $15,555.28 + $24,238.24 + $15,594.80 + $6,133.40 + $12,350.00 = $82,595.76

   Average stock:
   (6 BOM + 1 EOS) ÷ 7

   $60,280   (Round up to a whole number.)

   Turnover rate:
   Net sales ÷ average stock

   2.69   (Divide turnover rate to three places to the right of the decimal point and round to two places for more accuracy.)

3. Sales% each month:
Monthly sales ÷ total sales

15%, 13%, 12%, 20%, 30%, 10%

Stock-to-sales ratio:
BOM stock ÷ monthly sales

3.0, 2.6, 3.0, 3.1, 1.9, 4.0

Markdown% per month:
Monthly markdowns ÷ monthly sales

0.78%, 6%, 12.45%, 8.03%, 14.36%, 32.64%

Total markdowns ÷ total sales

11.57%

Monthly purchases at retail:

$35,250 + $143,820 + $355,450 + $149,450
+ $155,765 + $84,372 = $924,107

Monthly sales + EOM stock
+ monthly markdowns – BOM = purchases

Monthly purchases at cost:
100% – MU% = cost%
Retail$ × cost% = cost$

$17,272.50 + $70,471.80 + $174,170.50 +
$73,230.50 + $76,324.85 + $41,342.28 =
$452,812.43

Average stock:
(6 BOM + 1 EOS) ÷ 7

$398,457   (Round up to a whole number.)

Turnover rate:
Net sales ÷ average stock

2.26   (Divide turnover rate to three places to the
right of the decimal point and round to two
places for more accuracy.)

5. Sales% each month:
Monthly sales ÷ total sales

11.5%, 13.8%, 26.4%, 20.7%, 13.8%, 13.8%

Stock-to-sales ratio:
BOM stock ÷ monthly sales

3.1, 2.8, 2.0, 1.8, 1.8, 3.6

Markdown% per month:
Monthly markdowns ÷ monthly sales

10.01%, 5.08%, 5.67%, 8.94%, 16.03%, 18.78%

Total markdowns ÷ total sales

10%

Monthly purchases at retail:

$27,202 + $50,018 + $21,410 + $17,618 + $69,846
+ $17,306 = $203,400

Monthly sales + EOM stock
+ monthly markdowns – BOM = purchases

Monthly purchases at cost:
100% – MU% = cost%
Retail$ × cost% = cost$

$14,145.04 + $26,009.36 + $11,133.20 + $9,161.36
+ $36,319.92 + $8,999.12 = $105,768.00

Average stock:
(6 BOM + 1 EOS) ÷ 7

$69,771   (Round up to a whole number.)

Turnover rate:
Net sales ÷ average stock

2.49   (Divide turnover rate to three places to the
right of the decimal point and round to two places for
more accuracy.)

7. Sales% each month: 14.41%, 14.86%, 13.26%, 15.47%, 29.57%, 12.43%
   Monthly sales ÷ total sales

Due to rounding (because the percents are not exact), the sum of the percents may not be 100%. Therefore, you may have to round up instead of down or down instead of up to make sure the sum is exactly 100%.

Stock-to-sales ratio: 2.1, 2.3, 2.5, 2.8, 2.1, 2.7
BOM stock ÷ monthly sales

Markdown% per month: 6.31%, 9.69%, 14.29%, 8.77%, 7.18%, 18.29%
Monthly markdowns ÷ monthly sales

Total markdowns ÷ total sales   10%

Monthly purchases at retail: $126,845 + $100,725 + $166,925 + $234,850 +
                                         $19,500 + $142,000 = $790,845
Monthly sales + EOM stock
+ monthly markdowns – BOM = purchases

Monthly purchases at cost: $69,764.75 + $55,398.75 + $91,808.75 + $129,167.50
100% – MU% = cost%   + $10,725.00 + $78,100.00 = $434,964.75
Retail$ × cost% = cost$

Average stock: $260,533   (Round up to a whole number.)
(6 BOM + 1 EOS) ÷ 7

Turnover rate: 2.53   (Divide turnover rate to three places to the right
Net sales ÷ average stock   of the decimal point and round to two places for more
                                         accuracy.)

9. Sales% each month:
Monthly sales ÷ total sales

14.88%, 11.16%, 12.56%, 19.53%, 25.12%, 16.75%

Due to rounding (because the percents are not exact), the sum of the percents may not be 100%. Therefore, you may have to round up instead of down or down instead of up to make sure the sum is exactly 100%.

Stock-to-sales ratio:
BOM stock ÷ monthly sales

1.8, 2.0, 2.8, 2.8, 2.0, 2.5

Markdown% per month:
Monthly markdowns ÷ monthly sales

1.56%, 5.%, 5.56%, 13.1%, 13.3%, 11.1%

Total markdowns ÷ total sales

9.26%

Monthly purchases at retail:

$22,900 + $52,800 + $70,500 + $37,900 + $43,200
+ $52,000 = $279,300

Monthly sales + EOM stock
+ monthly markdowns – BOM = purchases

Monthly purchases at cost:
100% – MU% = cost%
Retail$ × cost% = cost$

$12,251.50 + $28,248.00 + $37,717.50 + $20,276.50
+ $23,112.00 + $27,820.00 = $149,425.50

Average stock:
(6 BOM + 1 EOS) ÷ 7

$85,543   (Round up to a whole number.)

Turnover rate:
Net sales ÷ average stock

2.51   (Divide turnover rate to three places to the right of the decimal point and round to two places for more accuracy.)

■ Chapter 9

1. Sales% each month:
   Monthly sales ÷ total sales

   13%, 24%, 20%, 17%, 16%, 10%

   Planned sales
   LY sales + % increase (decrease)
   LY total sales x monthly%

   $231,000
   210,000 + 10% = $231,000
   $30,030, $55,440, $46,200, $39,270, $36,960,
   $23,100

   Stock-to-sales ratio:
   BOM stock ÷ monthly sales

   3.07, 2.08, 2.55, 2.55, 2.7, 3.67

   Planned BOM stock
   Planned monthly sales x SSR

   $92,192, $115,315, $117,810, $100,139, $99,792,
   $84,777

   EOS%:
   LY EOS ÷ LY total sales

   44%

   Planned EOS:
   TY planned sales x EOS%

   $101,640

   Markdown% per month:
   Monthly markdowns ÷ monthly sales

   10.16%, 2.2%, 12.28%, 14.57%, 7.95%, 8.4%

   Planned markdowns:
   TY planned sales x monthly markdown %
   Total markdowns ÷ total sales

   $3,051 + $1,220 + $5,673 + $5,722 + $2,938 +
   $1,940 = $20,544
   8.89%

   Planned markdown%
   Total planned markdowns ÷ total planned
   sales

   8.89%

   Monthly purchases at retail:

   $56,204 + $59,155 + $34,202 + $44,645 + $24,883 +
   $41,903 = $260,992

   Planned monthly sales + planned EOM
   stock + planned monthly markdowns
   − planned BOM = purchases

   Monthly planned purchase at cost:
   100% − MU% = cost%
   Retail$ x cost% = cost$

   $29,788.12 + $31,352.15 + $18,126.06 + $23,661.85
   + $13,187.99 + $22,208.59 = $138,325.76

   Average stock:
   Planned (6 BOM + 1 EOS) ÷ 7

   $101,666   (Round up to a whole number.)

   Turnover rate:
   Net sales ÷ average stock

   2.27   (Divide turnover rate to three places to the
   right of the decimal point and round to two places for
   more accuracy.)

3. Sales% each month:
Monthly sales ÷ total sales

12%, 13%, 14%, 20%, 22%, 19%

Due to rounding (because the percents are not exact), the sum of the percents may not be 100%. Therefore, you may have to round up instead of down or down instead of up to make sure the sum is exactly 100%.

Planned sales
LY sales + % increase (decrease)
Planned total sales x monthly%

$329,000 − 10% = $296,100
$35,532, $38,493, $41,454, $59,220,
$65,142, $56,259

Stock-to-sales ratio:
BOM stock ÷ monthly sales

3.3, 3.2, 3.1, 2.9, 2.5, 2.2, 2.6

Planned BOM stock
Planned monthly sales x SSR

$117,256, $119,328, $120,217, $148,050,
$143,312, $146,273

EOS%:
LY EOS ÷ LY total sales

45.59%

Planned EOS:
TY planned sales x EOS%

$134,992

Markdown% per month:
Monthly markdowns ÷ monthly sales

4.17%, 6.23%, 12.14%, 6.0%, 13.73%,
14.74%

Planned markdowns
TY planned sales x monthly markdown %

$1,482 + $2,398 + $5,033 + $3,553 +
$8,944 + $8,293 = $29,703

Total markdowns ÷ total sales

10.03%

Planned markdown%
Total planned markdowns ÷ total planned
sales

10.03%

Monthly purchases at retail:

$39,086 + $41,780 + $74,320 + $58,035 +
$77,047 + $53,271 = $343,539

Planned monthly sales + planned EOM
stock + planned monthly markdowns
− planned BOM = purchases

Monthly planned purchase at cost:
100% − MU% = cost%
Retail$ x cost% = cost$

$22,669.88 + $24,232.40 + $43,105.60 +
$33,660.30 + $44,687.26 + $30,897.18 =
$199,252.62

Average stock:
Planned (6 BOM + 1 EOS) ÷ 7

$132,775   (Round up to a whole number.)

Turnover rate:
Net sales ÷ average stock

2.23   (Divide turnover rate to three places
to the right of the decimal point and round
to two places for more accuracy.)

5. Sales% each month:
   Monthly sales ÷ total sales

   13%, 24%, 20%, 17%, 16%, 10%

   Planned sales
   LY sales + % increase (decrease)
   LY total sales x monthly%

   $525,000 − 5% = $498,750
   $64,837, $119,700, $99,750, $84,788, $79,800, $49,875

   Note: Sales should be whole dollars, so you may need to round some values. In this case we rounded the first month down by 50¢ to $64,837; however, for November we rounded up to $84,788.

   Stock-to-sales ratio:
   BOM stock ÷ monthly sales

   1.88, 1.51, 1.89, 1.76, 1.87, 2.33

   Planned BOM stock
   Planned monthly sales x SSR

   $121,894, $180,747, $188,528, $149,227, $149,226, $116,209

   EOS%:
   LY EOS ÷ LY total sales

   30.67%

   Planned EOS:
   TY planned sales x EOS%

   $152,967

   Markdown% per month:
   Monthly markdowns ÷ monthly sales

   1.47%, 1.19%, 19.05%, 2.8%, 8.93%, 32.38%

   Planned markdowns:
   TY planned sales x monthly markdown %

   $953 + $1,424 + $19,002 + $2,374 + $7,126 + $16,150 = $47,029

   Total markdowns ÷ total sales

   9.43%

   Planned markdown%
   Total planned markdowns ÷ total planned sales

   9.43%

   Monthly purchases at retail:

   $124,643 + $128,905 + $79,451 + $87,161 + $53,909 + $102,783 = $576,852

   Planned monthly sales + planned EOM
   stock + planned monthly markdowns
   − planned BOM = purchases

   Monthly planned purchase at cost:
   100% − MU% = cost%
   Retail$ x cost% = cost$

   $72,292.94 + $74,764.90 + $46,081.58 + $50,553.38 + $31,267.22 + $59,614.14 = $334,574.16

   Average stock:
   Planned (6 BOM + 1 EOS) ÷ 7

   $151,257   (Round up to a whole number.)

   Turnover rate:
   Net sales ÷ average stock

   3.3   (Divide turnover rate to three places to the right of the decimal point and round to two places for more accuracy.)

■ Chapter 10

Answers will be different for every plan and for all students.

■ Chapter 11

1. $18,600 + $21,000 – $32,500 = $7,100

   $7,100 – $6,200 = $900   open to buy

3. $26,780 + $136,000 – $95,000 = $67,780

   $67,780 – $22,000 – $5,750 = $40,030

5. Open to buy is $63,500. It is already the middle of the month and a large amount of merchandise is needed to come in on stock plan. If under plan, sales can drop off without merchandise to sell.

7. Open to buy is $30,800.

   $44,000 + $56,000 + $800 = $100,800; $100,800 – $62,000 – $8,000 = $30,800

9. Overbought by $10,000.

   $32,000 + $48,000 + $6,500 – $65,000 = $21,500
   $31,500 – $21,500 = $10,000

   If the stock had come in on plan, they would still have been overbought by $2,000.

   $32,000 + $48,000 + $6,500 – $57,000 = $29,500
   $31,500 – $29,500 = $2,000

■ Chapter 12

1. Men's undershirts, classification 26
   Spring season, total BOM stock, $45,000
   Classification% to total stock, 35%; total planned dollars, $15,750.

Lower: $5.00, 20%, $3,150, 630 units

Middle: $8.00, 65%, $10,237.50, 1,280 units

Upper: $12.00, 15%, $2,362.50, 197 units

Lower: 50% round neck = 315 pcs/units
           50% V neck = 315 pcs/units

Middle: 50% round neck = 640 pcs/units
           50% V neck = 640 pcs/units

Upper: 50% round neck = 99 pcs/units
           50% V neck = 99 pcs/units

Lower

Small:	15%	47 round neck, 47 V neck
Medium:	30%	95 round neck, 95 V neck
Large:	40%	126 round neck, 126 V neck
XL:	15%	47 round neck, 47 V neck

Middle

Small:	15%	96 round neck, 96 V neck
Medium:	30%	192 round neck, 192 V neck
Large:	40%	256 round neck, 256 V neck
XL:	15%	96 round neck, 96 V neck

Upper (white)

			Upper (colors)		
Small:	15%	12 round neck, 2 V neck	Small:	15%	12 round neck, 2 V neck
Medium:	30%	24 round neck, 6 V neck	Medium:	30%	24 round neck, 6 V neck
Large:	40%	32 round neck, 8 V neck	Large:	40%	32 round neck, 8 V neck
XL:	15%	12 round neck, 3 V neck	XL:	15%	12 round neck, 2 V neck

3. Jogging shorts, classification 18
   May BOM, $385,000

   Classification% to total stock, 30%; total planned dollars, $115,500.

Lower: $16.00, 30%, $34,650, 2,165 units

Middle: $24.00, 60%, $69,300, 2,888 units

Upper: $32.00, 10%, $11,550, 361 units

Lower
Small:	15%	325 pieces
Medium:	20%	433 pieces
Large:	40%	866 pieces
XL:	25%	541 pieces

Middle
Small:	15%	433 pieces
Medium:	20%	578 pieces
Large:	40%	1,155 pieces
XL:	25%	722 pieces

Upper
Small:	15%	54 pieces
Medium:	20%	72 pieces
Large:	40%	145 pieces (rounded up)
XL:	25%	90 pieces

5. Boys 4–7 pants
 Fall total sales are $820,000.
 Classification% to total stock, 20%; total planned dollars, $164,000

 Casual pants

 70%; total planned dollars, $114,800
 Lower: $8.00, 20%, $22,960, 2,870 units
 Middle: $10.00, 50%, $57,440, 5,740 units
 Upper: $12.00, 30%, $34,440, 2,870 units

 Regulars: 60%, 1,722 pcs   Slim: 40%, 1,148 pcs

 $8.00
4	10%	172 pcs		4	0	0 pcs
5	20%	345 pcs	(rounded)	5	30%	344 pcs
6	40%	689 pcs		6	50%	574 pcs
7	30%	516 pcs		7	20%	230 pcs (rounded up)

 Regulars: 60%, 3,444 pcs   Slim: 40%, 2,296 pcs

 $10.00
4	10%	344 pcs		4	0	0 pcs
5	20%	689 pcs	(rounded)	5	30%	689 pcs
6	40%	1378 pcs		6	50%	1148 pcs
7	30%	1033 pcs		7	20%	459 pcs (rounded up)

 Regulars: 60%, 1,722 pcs   Slim: 40%, 1,148 pcs

 $12.00
4	10%	172 pcs		4	0	0 pcs
5	20%	345 pcs	(rounded)	5	30%	344 pcs
6	40%	689 pcs		6	50%	574 pcs
7	30%	516 pcs		7	20%	230 pcs (rounded up)

Dressy pants

30%; total planned dollars, $49,200
Lower: $10.00, 30%, $14,760, 1,476 units
Middle: $12.00, 50%, $24,600, 2,050 units
Upper: $15.00, 20%, $9,840, 6,560 units

Regulars: 60%, 886 pcs   Slim: 40%, 590 pcs

$10.00

4	10%	89 pcs		4	0	0 pcs
5	20%	177 pcs	(rounded)	5	30%	177 pcs
6	40%	354 pcs		6	50%	295 pcs
7	30%	266 pcs		7	20%	118 pcs (rounded up)

$12.00

4	10%	123 pcs	4	0	0 pcs
5	20%	246 pcs	5	30%	246 pcs
6	40%	492 pcs	6	50%	410 pcs
7	30%	396 pcs	7	20%	164 pcs

$15.00

4	10%	394 pcs	4	0	0 pcs
5	20%	787 pcs	5	30%	787 pcs
6	40%	1,574 pcs	6	50%	1,312 pcs
7	30%	1,181 pcs	7	20%	525 pcs

■ Chapter 13

1.

Net sales	$35,000	100	%
– CMS	$18,000	51.43%	
Gross margin	$17,000	48.57%	
– Operating expenses	$ 7,200	20.57%	
Profit (or loss)	$ 9,800	28	%

3.

Net sales	$400,000	100	%
– CMS	$253,000	63.25%	
Gross margin	$147,000	36.75%	
– Operating expenses	$149,000	37.25%	
Profit (or loss)	($2,000)	(0.5%)	

5.

Net sales	$141,000	100 %
− CMS	$ 87,000	61.70%
Gross margin	$ 54,000	38.30%
− Operating expenses	$ 27,500	19.50%
Profit (or loss)	$ 26,500	18.8 %

7. $7,000 − $80 + $100 + $90 = $7,110   net cost of merchandise

9. $14,280 − $7,440 = $6,840   gross margin

11. $100,000 − 8% + $1,502 = $93,502
$140,000 − $93,502 = $46,498   gross margin

13.

Gross sales	$909,290	
− CRA	$ 51,466	
Net sales	$857,824	100 %
− CGS	$471,342	54.95%
Gross margin	$386,482	45.05%

Direct expenses

Buying salary	$ 9,300
Selling salaries	$55,170
Dept. advertising	$24,876
Dept. rec. exp.	$ 7,145

Total direct expenses	$ 96,491	11.25%
Contribution Margin	$289,991	33.80%

Indirect expenses

Allocated:

Rent	$25,360
Exec. salary	$39,060
Utilities	$25,430
Maintenance	$16,880

Total indirect expenses	$106,730	12.44%
Net profit	$183,261	21.36%

15.

Gross sales		$102,500	
− CRA		$ 15,000	
Net sales		$ 87,500	100%

Opening inventory	$39,000		
Purchases at cost	$42,400		
Transportation	$ 1,250		
− Closing Inventory	$36,750		
− Cash Discounts	$ 1,200		
Net GS		$44,700	51.09%
Gross margin		$42,800	48.91%

Direct expenses			
Buying salary	$4,500		
Selling salaries	$5,450		
Dept. advertising	$6,500		
Total direct expenses		$16,450	18.80%

Contribution margin		$26,350	30.11%

Indirect expenses			
Allocated:			
Rent	$ 5,500		
Exec. salary	$ 7,250		
Maintenance	$10,000		
Total indirect expenses		$22,750	26.00%
Net profit		$ 3,600	4.11%

■ Chapter 14

1. 12 × 8 = 96 sq. ft
$82,500 ÷ 96 = $859.38 annual sales per square foot

3. 22 × 44 = 968 sq. ft
   $178,500 ÷ 968 = $184.40

5. 35 × 15 = 525 sq. ft
   1,568,000 × 11% = $172,480
   $172,480 ÷ 525 = $328.53

7. 65 × 58 = 3,770 sq. ft overall
   10 × 4 = 40 sq. ft of stock area
   3 × 29 = 87 sq. ft dressing room and checkout area
   3 × 20 = 60 sq. ft for display area

   3,770 – 40 – 87 – 60 = 3,583 sq. ft overall

   $1,542,000 ÷ 3,583 = $430.37 annual sales per square foot

   $430.37 ÷ 12 = $35.86

9. 20 × 18 = 360 sq. ft overall for the men's shoe division
   60 × 65 = 3,900 sq. ft for the department overall
   10 × 2 = 20 sq. ft of stock area
   360 ÷ 3,900 = 9.23%
   360 – 20 = 340 sq. ft of selling space
   $152,000 ÷ 340 = $447.06

   $912,000 ÷ 3,900 = $233.85 sales per square foot for the entire department for the season. Productivity for the men's area is almost double, with less than 10% of the overall square footage. Discuss what might be the outcome of expanding the men's area. What would be some of the areas you would want to research? For example, where is the men's area on the floor? What type of fixtures are there? What are the sales trends? Are there special promotions?

■ Chapter 15

1. a.

COST SHEET

Date _August 8_

Style number _875_

Sketch and materials swatch

Description _Spring bridesmaid dress_

Fabric mill _Taiwan_

Manufactured _Hong Kong_

Size range _6, 8, 10, 12, 14_

Size scale _1-2-3-3-3_

Fabric width _36 in._

Estimated marker yardage _39 yd_

Actual marker yardage _n/a_

Materials	Yardage	Price per Yard	Total Amount
Fabric	3.25	$2.48	$8.06
Fabric (collar)			
Lining	0.75	$1.29	$0.97
		Total cost of materials	$9.03

Trims	Amounts	Cost Each	Total Amount
Buttons	1	$0.35	$0.35
Zippers			
Belts			
Other (elastic tape)	4 yd	$0.67	$2.68
	Total cost of trims/findings		$ 3.03

Labor

Marking costs (combined)   Grading costs _$0.57_   Cutting costs _$0.35_

Construction _$7.00_   Total labor costs   _$7.92_

Shipping and Duty Expenses

Import duties _(Combined)_   Shipping expenses _$1.12_

Total shipping and duty costs   _$1.12_

Total cost of goods to be sold _$21.10_

Suggested wholesale price _$52.75_

Suggested retail price _$105.50_

b.

	Dollars	Percents
Wholesale price	$52.75	100%
Markup%	$31.65	60%
Cost of goods sold	$21.10	40%

$21.10 ÷ 40% = $52.75 wholesale price
$52.75 − $21.10 = $31.65 markup dollars

c.

	Dollars	Percents
Retail selling price	$105.50	100%
Markup%	$52.75	50%
Cost/wholesale price	$52.75	50%

The retail price is $105.50. The buyer might round up to $106.00 or down to $105.00 to maintain a whole number for the retail selling price.

3. a.

COST SHEET

Date _May 14_

Style number _1890_

Sketch and materials swatch

Description _Jogging shorts_

Fabric mill _Honduras_

Manufactured _Honduras_

Size range _S—M—L_

Size scale _2—4—6_

Fabric width _45 in_

Estimated marker yardage _7.5_

Actual marker yardage _n/a_

Materials	Yardage	Price per Yard	Total Amount
Fabric	0.625	$1.45	$0.91
Fabric (collar)			
Lining			
		Total cost of materials	$0.91

Trims	Amounts	Cost Each	Total Amount
Buttons			
Zippers			
Belts			
Other (elastic tape)	0.75	$0.65	$0.49
		Total cost of trims/findings	$0.49

Labor

Marking costs (combined)   Grading costs _$0.25_   Cutting costs _$0.45_

Construction _$1.50_   Total labor costs   _$2.20_

Shipping and Duty Expenses

Import duties _n/a_   Shipping expenses _$0.82_

Total shipping and duty costs   _$0.82_

Total cost of goods to be sold _$4.42_

Suggested wholesale price _$6.31_

Suggested retail price _$12.00_

b.

	Dollars	Percents
Wholesale price	$6.31	100%
Markup	$1.89	30%
Cost of goods sold	$4.42	70%

$4.42 ÷ 70% = $6.31 suggested wholesale price.

c.

	Dollars	Percents
Retail selling price	$11.82	100%
Markup	$ 5.32	45%
Cost/wholesale price	$ 6.50	55%

These jogging shorts will retail for $12.00.

e.

	Dollars	Percents
Retail selling price	$9.74	100%
Markup	$5.32	54.62%
CGS	$4.42	45.38%

Cost of goods to be sold was $4.42. That will never change. The markup earned by the retail company was $5.32. If a private label was developed for a company, the same product could retail for $9.74 and the company would earn $5.32 markup dollars, which is 54.62%. If the buyer marked the shorts $10.00, the retail markup would be $5.58, which would be 55.8%—more profit but a lower retail price. Look how much lower the retail selling price is, even though your company is earning the same retail markup dollars and the cost of the goods sold is the same.

5.

	Dollars	Percents
Cost/wholesale	$7.27	100%
Markup	$3.27	45%
CGS	$4.00	55%

$100\% - 45\% = 55\%$

$\$4.00 \div 55\% = \$7.27$

Set the cost/wholesale price at $7.50.

	Dollars	Percents
Retail selling price	$15.00	100%
Markup	$ 7.50	50%
Cost/wholesale price	$ 7.50	50%

index